POLYMERS AND POLYMERIC COMPOSITES

Properties, Optimization, and Applications

AAP Research Notes on
Polymer Engineering Science and Technology

POLYMERS AND POLYMERIC COMPOSITES

Properties, Optimization, and Applications

Edited by
Liliya I. Bazylak, PhD

Gennady E. Zaikov, DSc, and A. K. Haghi, PhD
Reviewers and Advisory Board Members

Apple Academic Press
TORONTO NEW JERSEY

Apple Academic Press Inc.
3333 Mistwell Crescent
Oakville, ON L6L 0A2
Canada

Apple Academic Press Inc.
9 Spinnaker Way
Waretown, NJ 08758
USA

International Standard Book Number-13: 978-1-77188-049-7 (hbk)

Library of Congress Control Number: 2014954052

Library and Archives Canada Cataloguing in Publication

Polymers and polymeric composites: properties, optimization, and applications/edited by Liliya I. Bazylak, PhD; Gennady E. Zaikov, DSc, and A.K. Haghi, PhD, Reviewers and Advisory Board Members.

(AAP research notes on polymer engineering science and technology series)
Includes bibliographical references and index.
ISBN 978-1-77188-049-7 (bound)
1. Polymers. 2. Polymeric composites. I. Bazylak, Liliya I., editor
II. Series: AAP research notes on polymer engineering science and technology series

TA455.P58P68 2014 620.1'92 C2014-907146-9

Apple Academic Press also publishes its books in a variety of electronic formats. Some content that appears in print may not be available in electronic format. For information about Apple Academic Press products, visit our website at **www.appleacademicpress.com** and the CRC Press website at **www.crcpress.com**

ABOUT AAP RESEARCH NOTES ON POLYMER ENGINEERING SCIENCE AND TECHNOLOGY

The AAP Research Notes on Polymer Engineering Science and Technology reports on research development in different fields for academic institutes and industrial sectors interested in polymer engineering science and technology. The main objective of this series is to report research progress in this rapidly growing field.

Alfonso Jimenez, PhD
Professor of Analytical Chemistry and Materials Science, University of Alicante, Spain

Gennady E. Zaikov, DSc
Head, Polymer Division, N. M. Emanuel Institute of Biochemical Physics, Russian Academy of Sciences; Professor, Moscow State Academy of Fine Chemical Technology, Russia; Professor, Kazan National Research Technological University, Kazan, Russia

BOOKS IN THE AAP RESEARCH NOTES ON POLYMER ENGINEERING SCIENCE AND TECHNOLOGY SERIES

Functional Polymer Blends and Nanocomposites: A Practical Engineering Approach
Editors: Gennady E. Zaikov, DSc, Liliya I. Bazylak, PhD, and A. K. Haghi, PhD

Polymer Surfaces and Interfaces: Acid-Base Interactions and Adhesion in Polymer-Metal Systems
Irina A. Starostina, DSc, Oleg V. Stoyanov, DSc, and Rustam Ya. Deberdeev, DSc

Key Technologies in Polymer Chemistry
Editors: Nikolay D. Morozkin, DSc, Vadim P. Zakharov, DSc, and Gennady E. Zaikov, DSc

Polymers and Polymeric Composites: Properties, Optimization, and Applications
Editors: Liliya I. Bazylak, PhD, Gennady E. Zaikov, DSc, and A. K. Haghi, PhD

ABOUT THE EDITOR

Liliya I. Bazylak, PhD

Senior Staff Scientist, Physical-Chemistry of Combustible Minerals Department, Institute of Physical Organic Chemistry and Coal Chemistry, National Academy of Sciences of Ukraine, Lviv, Ukraine

Liliya Bazylak, PhD, is Senior Staff Scientist in the Physical and Chemistry of Combustible Minerals Department at the Institute of Physical and Organic Chemistry and Coal Chemistry at the National Academy of Sciences of Ukraine in Lviv. She is the author of more than 200 publications. Her scientific interests include physical chemistry, nanochemistry and nanotechnologies, and also chemistry of high molecular compounds.

REVIEWERS AND ADVISORY BOARD MEMBERS

Gennady E. Zaikov, DSc

Gennady E. Zaikov, DSc, is Head of the Polymer Division at the N. M. Emanuel Institute of Biochemical Physics, Russian Academy of Sciences, Moscow, Russia, and Professor at Moscow State Academy of Fine Chemical Technology, Russia, as well as Professor at Kazan National Research Technological University, Kazan, Russia. He is also a prolific author, researcher, and lecturer. He has received several awards for his work, including the Russian Federation Scholarship for Outstanding Scientists. He has been a member of many professional organizations and on the editorial boards of many international science journals.

A. K. Haghi, PhD

A. K. Haghi, PhD, holds a BSc in urban and environmental engineering from University of North Carolina (USA); a MSc in mechanical engineering from North Carolina A&T State University (USA); a DEA in applied mechanics, acoustics and materials from Université de Technologie de Compiègne (France); and a PhD in engineering sciences from Université de Franche-Comté (France). He is the author and editor of 65 books as well as 1000 published papers in various journals and conference proceedings. Dr. Haghi has received several grants, consulted for a number of major corporations, and is a frequent speaker to national and international audiences. Since 1983, he served as a professor at several universities. He is currently Editor-in-Chief of the *International Journal of Chemoinformatics and Chemical Engineering* and *Polymers Research Journal* and on the editorial boards of many international journals. He is a member of the Canadian Research and Development Center of Sciences and Cultures (CRDCSC), Montreal, Quebec, Canada.

CONTENTS

LIST OF CONTRIBUTORS

Esen Arkış
Izmir Institute of Technology Department of Chemical Engineering, Gülbahce Urla 35430 Izmir Turkey

M. I. Artsis
Institute of Biochemical Physics after N. M. Emanuel, Kosygina 4, Moscow, 119991, Russia

D. V. Bagrov
Faculty of Biology, Moscow State University, Leninskie gory 1-12, 119992 Moscow, Russia

Devrim Balköse
Izmir Institute of Technology Department of Chemical Engineering, Gülbahce Urla 35430 Izmir Turkey

A. P. Bonartsev
A.N. Bach's Institute of Biochemistry, Russian Academy of Sciences, Leninskiy prosp. 33, 119071 Moscow, Russia, Faculty of Biology, Moscow State University, Leninskie gory 1-12, 119992 Moscow, Russia

G. A. Bonartseva
A.N. Bach's Institute of Biochemistry, Russian Academy of Sciences, Leninskiy prosp. 33, 119071 Moscow, Russia

A. P. Boskhomodgiev
A.N. Bach's Institute of Biochemistry, Russian Academy of Sciences, Leninskiy prosp. 33, 119071 Moscow, Russia

Jakub Czakaj
AIB Ślączka, Szpura, Dytko spółka jawna, Knurów

E. A. Filatova
A. N. Bach's Institute of Biochemistry, Russian Academy of Sciences, Leninskiy prosp. 33, 119071 Moscow, Russia

A. L. Iordanskii
A.N. Bach's Institute of Biochemistry, Russian Academy of Sciences, Leninskiy prosp. 33, 119071 Moscow, Russia, N.N. Semenov Institute of Chemical Physics, Russian Academy of Sciences, Kosygin str. 4, 119991 Moscow, Russia. Tel.: +74959397434; Fax: +74959382956; E-mail: aljordan08@gmail.com

Aleksei A. Iordanskii
Semenov Institute of Chemical Physics, Russian Academy of Sciences, Moscow, Russia

E. A. Ivanov
A. N. Bach's Institute of Biochemistry, Russian Academy of Sciences, Leninskiy prosp. 33, 119071 Moscow, Russia

Svetlana G. Karpova
Emmanuel Institute of Biochemical Physics, Russian Academy of Sciences, Moscow, Russia Leninskii pr., 32 a, Moscow 119991, Russian Federation

G. V. Kozlov
Institute of Applied Mechanics of Russian Academy of Sciences, Russia

Sergei M. Lomakin
Institute of Organic Chemistry, Ufa Scientific Center of Russian Academy of Sciences, pr. Oktyabrya 71, Ufa, Bashkortostan, 450054, Russia

T. K. Makhina
A.N. Bach's Institute of Biochemistry, Russian Academy of Sciences, Leninskiy prosp. 33, 119071 Moscow, Russia

V. L. Myshkina
A. N. Bach's Institute of Biochemistry, Russian Academy of Sciences, Leninskiy prosp 33, 119071 Moscow, Russia

Anatolii A. Popov
Bashkir State University, Zaki Validi str. 32, Ufa, 450076 Bashkortostan, Russia

Maria Rajkiewicz
Institute for Engineering of Polymer Materials and Dyes, Department of Elastomers and Rubber Technology in Piastów

A. V. Rebrov
A.V. Topchiev Institute of Petroleum Chemistry. Leninskiy prosp, 27, 119071 Moscow, Russia

Marcin Ślączka
AIB Ślączka, Szpura, Dytko spółka jawna, Knurów

S. A. Yakovlev
A.N. Bach's Institute of Biochemistry, Russian Academy of Sciences, Leninskiy prosp. 33, 119071 Moscow, Russia

Yu. G. Yanovskii
Institute of Applied Mechanics of Russian Academy of Sciences, Leninskii pr., 32 a, Moscow 119991, Russian Federation

G. E. Zaikov
N. M. Emanuel Institute of Biochemical Physics of Russian Academy of Sciences, Kosyginst., 4, Moscow 119334/119991, Russia

LIST OF ABBREVIATIONS

AFM	Atomic Force Microscopy
AP	Aromatic Polyesters
BSR/TC	Butadiene-Styrene Rubber/Technical Carbon
CNT	Carbon Nano Tube
DBTL	Dibutyl Tin Dilaurate
DFT	Density Function Method
DMTA	Dynamic Thermal Analysis of Mechanical Properties
DPD	Dissipative Particle Dynamics
DSC	Differential Scanning Calorimetry
DWNTs	Double Walled Carbon Nanotubes
EOE	Ethylene-Octene Elastomer
EPDM	Ethylene-Propylene-Diene Elastomer
GSDBT	Generalized Shear Deformation Beam Theory
HH	3-hydroxyheptanoate
HO	3-hydroxyoctanoate
HR	Heat Radiation
HV	3-hydroxyvalerate
iPP	Isotactic Polypropylene
LB	Lattice Boltzmann
LD	Local Density
LQPs	Liquid-Crystal Polyesters
MC	Monte Carlo
MD	Molecular Dynamics
MW	Molecular Weight
MWNTs	Multi-Walled Nanotubes
PHAs	Polyhydroxyalkanoates
PHB	Poly-3-Hydroxybutyrate
PIB	Polyisobutylene
PSDT	Parabolic Shear Deformation Theory
QC	Quasi-Continuum

QM	Quantum Mechanics
RBM	Radial Breathing Mode
RHR	Rate of Heat Release
RVE	Representative Volume Element
SWNTs	Single Walled Nanotubes
TBMD	Tight Bonding Molecular Dynamics
TGA	Thermogravimetric Analysis
TPE	Thermoplastic Elastomers
TPE-V	Thermoplastic Vulcanisates
UC	University of Cincinnati
VIM	Variational Iteration Method
WAXS	Wide Angle X-Ray Scattering

LIST OF SYMBOLS

a	lower linear scale of fractal behavior
c	nanoparticles concentration
C_∞	characteristic ratio
d	dimension of Euclidean space
k_n	proportionality coefficient
K_B	Boltzmann constant
K_T	isothermal modulus of dilatation
l_0	chain skeletal bond length
M	current molecular weight
N	number of particles
S	macromolecule cross-sectional area
S_i	quadrate area
S_n	cross-sectional area of nanoparticles
S_u	nanoshungite particles
T	temperature
t	walk duration
T_g	glass transition temperature
W_n	nanofiller mass content

Greek Symbols

α	numerical coefficient
β	probability of chain termination
η	medium viscosity
ν	Poisson ratio
$\psi(\beta)$	distribution of active site over kinetic heterogeneity
φ_{if}	interfacial regions relative fraction
ρ_n	density

PREFACE

The main attention in this collection of scientific papers is on the recent theoretical and practical advances in polyblends and composites. This volume highlights the latest developments and trends in advanced polyblends and their structures. It presents new developments of advanced polyblends and respective tools to characterize and predict the material properties and behavior. The book provides important original and theoretical experimental results, which use nonroutine methodologies often unfamiliar to the usual readers. Chapters in this book also present novel applications of more familiar experimental techniques and analyses of composite problems which indicate the need for new experimental approaches presented.

Technical and technological developments demand the creation of new materials, that are stronger, more reliable, and more durable—that is, materials with new properties. Up-to-date projects in creation of new materials go along the way of nanotechnology.

The technology of polyblends manufacturing forges ahead; its development is directed to the simplification and cheapening of the production processes of composite materials with nanoparticles in their structure. However, new nanotechnologies develop very quickly; what seemed impossible yesterday will be accessible to the introduction on a commercial scale tomorrow. The desired event of fast implementation of polyblends in mass production depends on the efficiency of cooperation between the scientists and the manufacturers in many respects. Today's high technology problems of applied character are successfully solved with the coordinated efforts of both the scientific and business worlds.

With contributions from experts from both the industry and academia, this book presents the latest developments in the identified areas. This book incorporates appropriate case studies, explanatory notes, and schematics for clarity and understanding.

This book will be useful for chemists, chemical engineers, technologists, and students interested in advanced nano-polymers with complex behavior and their applications.

CHAPTER 1

A STUDY ON ABSORPTION AND REFLECTION OF INFRARED LIGHT BY THE UNCOATED AND AL COATED SURFACES OF POLYMER FILMS TECHNIQUES

ESEN ARKIŞ and DEVRIM BALKÖSE

CONTENTS

ABSTRACT

Polymer films coated with a thin layer of aluminum or aluminum oxide are extensively used in food packing as heat shields. The infrared rays were not transmitted through the films and were reflected protecting the contents from the harmful effects of infrared light. The quantitative measurement of the film thickness and infrared light reflection and absorption capacities of aluminum coated films used as packing materials were possible using infrared spectroscopy.

1.1 INTRODUCTION

The protection from harmful effects caused by infrared radiation on foods can be made by using infrared shielding packing materials. Nano-oxides with the surface effect, small size effect, quantum size effect, macroscopic quantum tunneling effect and other special properties can be used in preparation of infrared shielding coating, absorbing coatings, conductive coatings, insulation coatings and so on [1].

Polymer coatings on the surfaces are very easily degraded because of the infrared light. The life of the polymer coating could be extended if they are coated with an infrared light reflecting or absorbing layer. Aluminum coatings were used for this purpose in many applications [h2]. Aluminum oxide coating on polymers is also used for protection of the polymer layer. Covalent bonds are responsible for adhesion of aluminum oxide to polymer surface. Nano Al_2O_3 particles have a wide absorption band in the infrared band and they have been used widely in the paint, military, scientific and industrial application [2]. Resonance absorption of the infrared radiation may take place at extremely thin metal coatings by reducing the overall temperature of the heat shield [3].

Polymer surfaces are modified by plasma techniques for interfacial enhancement [4–10]. Au, Ag, Pd, Cu and Ni were coated on poly (methylmethacrylate) (PMMA) by barrel technique [11]. Coating of Al alloys on PET was compared with Ti layer under Al alloys [12]. Thin Aluminum oxide coatings have been deposited on various uncoated papers, polymer-coated papers and plain polymer films using atomic layer deposition technique [13]. The isotactic polypropylene (iPP) and Al composite is widely used as television cable electromagnetic shielding materials [14]. The reflection of infrared light depends on geometry of the surface of aluminum. The reflection of infrared light was reduced when the height of the triangular aluminum gratings were reduced [15]. Al coating obtained by roll-to-roll coating on polypropylene was polycrystalline with a grain size 20–70 nm [16].

The morphology, order, light transmittance and water vapor permeability of the Al coated polypropylene films were reported in a previous publication [17]. The films did not transmit light in the UV and visible region of the light spectrum and Al coating reduced the water vapor permeation through the films [17].

The objective of this chapter is to study the absorption and reflection of infrared light by the uncoated and Al coated surfaces of polymer films using transmission and reflection techniques. For this purpose two commercial biaxially oriented polypropylene film, a cast polypropylene film, a milk cover and a chocolate coating material with their one surface coated with a thin aluminum layer were examined by infrared spectroscopy. One surface of the each film was coated with a thin aluminum layer.

1.2 EXPERIMENTAL PART

1.2.1 MATERIALS

Infrared light absorption and reflection of the uncoated and Al coated surfaces of polymer films were investigated and discussed in this chapter. Two commercial biaxially oriented polypropylene samples coated with Al by physical vapor deposition technique were examined. The samples called commercial film 1 and commercial film 2 were provided by POLINAS and POLIBAK companies, respectively. They were 16 μm and 19 μm thick and their one surface was coated with Al after a corona discharge treatment.

An aluminum coated polypropylene film with 50-μm thickness that was previously prepared Ozmihci et al. [17] was also examined. Al film on polypropylene was deposited by the high vacuum magnetron sputtering system having four guns. To create plasma Argon gas (99.9 wt.%) was used. Polypropylene film fixed on glass substrates was coated by condensation of Al atoms sputtered from an Al target. In the sputtering process 20 W DC power and 20 mA current were applied to pure Al target. The Al coating obtained by magnetron sputtering method was 98–131 nm thickness and formed by small Al particles having 22–29 nm grain sizes [18]. Commercial milk and chocolate packaging materials were also investigated.

1.2.2 METHODS

The surface morphology of the films was examined by scanning electron microscopy. FEI QUANTA FEG-250 SEM was used for this purpose. The thickness of Al coating on the surface of the films was measured from the fractured brittle Al layer obtained by stretching the films.

The transmission infrared spectra of the films at 20 °C were taken with Excalibur DIGILAB FTS 3000 MX type Fourier Transform Infrared Spectrophotometer with a resolution of 4 cm^{-1}. DTGS type detector was used for all measurements. The transmission spectra of the films were obtained by placing the films in two different positions. The incident infrared light first strikes to either Al coated or uncoated surface in these positions. In transmission spectrum the light that passes through the sample is measured. The grazing angle specular reflectance accessory with 80 °C (Pike Technologies), in the reflection absorption mode was used to obtain specular reflectance spectra. Gold-coated glass was used as the reference.

1.3 RESULTS AND DISCUSSION

1.3.1 MORPHOLOGIES OF AL COATINGS

SEM micrographs of the Al coated surfaces of the films were taken to observe the morphology of the coatings on the films. The Al coating on the surface of the films were broken by stretching the polymer phase to observe the thickness of Al coatings.

1.3.1.1 MORPHOLOGY OF COMMERCIAL FILM 1

The commercial polypropylene films were coated with a layer of Al by physical vapor deposition. The coating thickness was determined from SEM pictures of commercial film 1 seen in Fig. 1.1a as 32 nm for commercial film 1. The coating had the shape of the polymer layer. There were parallel lines on the surface of the polymer produced during processing of the polypropylene film in continuous film machinery.

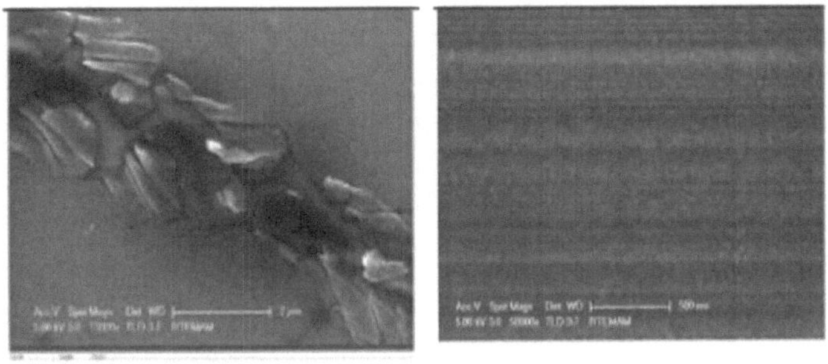

(a) (b)

FIGURE 1.1 SEM pictures of Al coated surface of commercial film 1. (a) the Al surface fractured to observe coating thickness; (b) the surface as produced.

1.3.1.2 MORPHOLOGY OF COMMERCIAL FILM 2

In Fig. 1.2, SEM micrograph of commercial Film2 is shown. Figure 1.2a is fractured Al coating on the surface. Coating thickness is determined as 185 nm from (Fig. 1.2a). Figure 1.2b is Al coated polypropylene surface. The coating was made up of Al particles condensed on the surface of the polypropylene phase. There were uncoated regions on the surface.

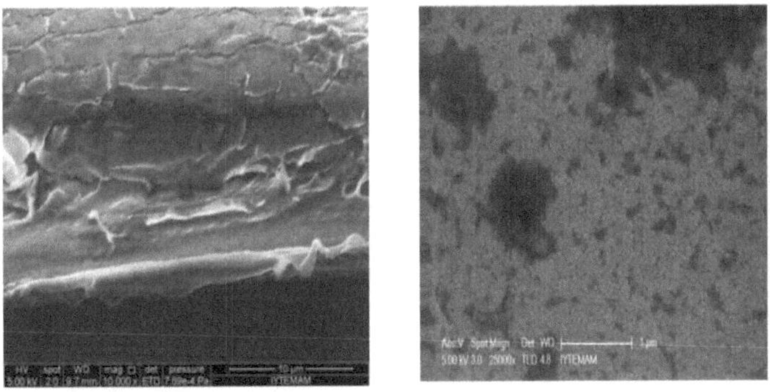

(a) (b)

FIGURE 1.2 SEM Pictures from BOPP Metalized commercial film 2. (a) Fractured Al Surface; (b) Coated polypropylene.

1.3.1.3 MORPHOLOGY OF MAGNETRON SPUTTERED FILMS

The polypropylene films were at 50-μm thickness and their one surface was covered with 98–131 nm thick Al layer with magnetron sputtering [18]. In Fig. 1.3, SEM micrographs of magnetron sputtered film are shown. In Fig. 1.3a, there is fractured Al coating on the surface of polypropylene. Figure 6 b. is the top view of the Al layer. From Fig. 1.3a, the coating thickness is determined to be 185 nm, close to the value reported previously [18]. The top view of the Al layer seen in Fig. 1.3b indicated the surface was covered by Al particles having nearly 30 nm size. The surface was not very smooth. There are cracks on the surface of the brittle Al coating.

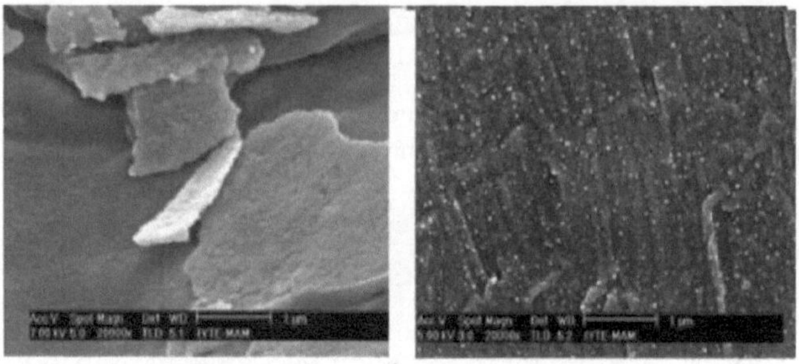

(a) (b)

FIGURE 1.3 SEM pictures of (a) fractured Al coating on the cast polypropylene film; (b) Al coated surface of the polypropylene film.

1.3.1.4 MORPHOLOGY OF MILK AND CHOCOLATE PACKAGES

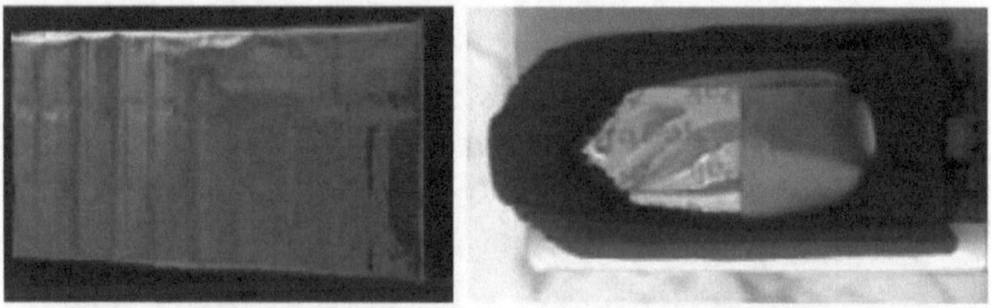

(a) (b)

FIGURE 1.4 The photographs of (a) Chocolate packing (b) Milk cover.

The chocolate packing had an outer layer coated with aluminum. The inner layer was Milk cover had the function of opening the packing when it was pulled. The inner side of the cover, which was in contact with the milk, was covered with a polymer layer.

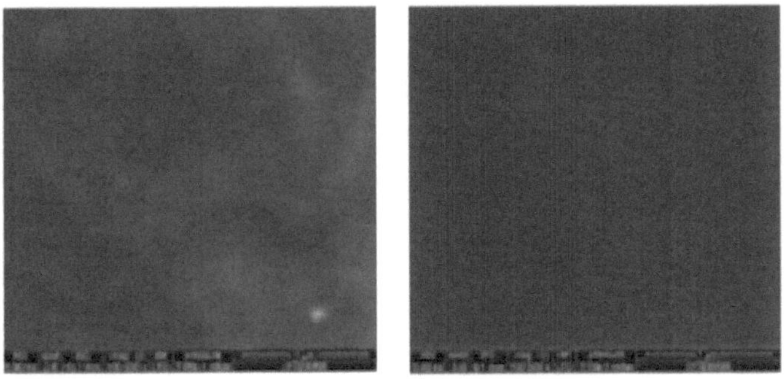

(a) (b)

FIGURE 1.5 The SEM micrographs Al coated surface of (a) the chocolate packing; (b) milk cover.

Photographs of chocolate packing and milk cover are shown in Fig. 1.4. The SEM micrographs of their Al coated surfaces are in Fig. 1.5. They have an even Al coating as seen in Fig. 1.5.

1.3.2 TRANSMISSION SPECTRA OF AL COATED POLYPROPYLENE FILMS

1.3.2.1. COMMERCIAL AL COATED POLYPROPYLENE FILMS

The transmission infrared spectra of commercial film 1 and 2 were obtained by placing the films in the path of the infrared light. Spectrum 1 and spectrum 2 in Fig. 1.6 belonged to commercial film 1 and film 2, respectively. The base line absorbance values were close to 1for commercial film 1 and to 4 for commercial film 2.This is that indicated 1/10,000 and 1/10 of the input infrared light was transmitted from the samples. The Al coating on the film acted as a barrier to infrared light. The higher coating thickness of commercial film 2 (185 nm) than that of commercial film 1 (30 nm) resulted higher level of shielding from infrared radiation. The film 2 had better infrared light shielding efficiency than the film 1.

The peaks observed at 3000–2800 cm^{-1} belonged to C–H asymmetric and symmetric stretching, at 1450 cm^{-1} C–H bending, at 1350 cm^{-1} C–H deformation bending [17]. The peak at 973 cm^{-1} belongs to amorphous CH_3rocking and C–C chain stretching vibrations and the peak at 998 cm^{-1} belongs to crystalline CH_3 rocking, CH_2 wagging and CH bending vibrations [20].

The magnetron sputtered film's polymer layer and Al layer were 50 μm and 185 nm, respectively. This film was nearly three times thicker than commercial film 1 and commercial film 2. The absorbance values of this film in transmission mode were too high, since both the Al layer and polypropylene layer were thicker. The FTIR transmission spectrum of this thick film coated by magnetron sputtering had very high absorbance values indicating it was also a good shield for infrared radiation.

1.3.3 SPECULAR REFLECTANCE OF SPECTRA OF THE FILMS

1.3.3.1 COMMERCIAL AL COATED POLYPROPYLENE FILM 1

Figure 1.6 shows the specular reflectance spectra of Al coated and uncoated surfaces of the film 1. Al coated surface reflected 100% of the infrared rays since the absorbance values were very close to zero. Uncoated surface had the characteristic spectrum of polypropylene. However, there were other peaks observed called fringes due to reflection of light from both surfaces of the thin film.

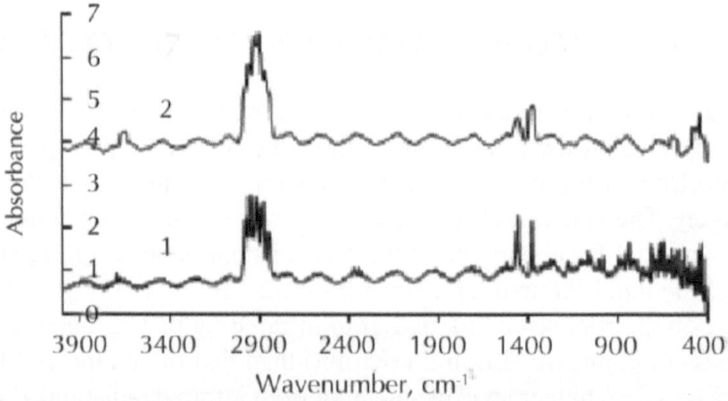

FIGURE 1.6 Transmittance spectra of the commercial film 1 and 2.

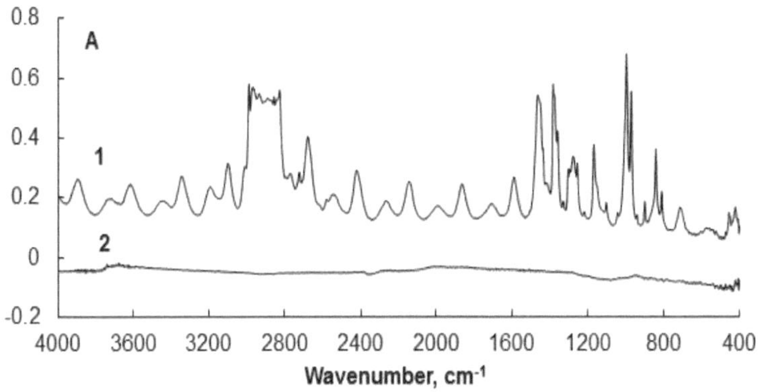

FIGURE 1.7 Specular Reflectance of Spectra of the film 1 (1) polypropylene surface (2) Al coated surface of the film 1.

1.3.3.2 THICKNESS OF COMMERCIAL FILM 1

From the number of the fringes the thickness of the films can be calculated using Eq. (1) [19].

$$b = \frac{1}{2(n)} x \frac{N}{(\nu_1 - \nu_2)} \tag{1}$$

where b = film thickness, n = refractive index of sample, N = number of fringes within a given spectral region, ν_1, ν_2 = start and end point in the spectrum in cm^{-1}.

To count the number of fringes, select starting and ending points both as minima or maxima of the spectrum and then count the number of opposing minima or maxima. In other words if we select minima values for starting and ending points in the spectrum, then select maxima points to count the number of fringes.

The refractive index of polypropylene is 1.49. There are 5 refraction fringes in the region of 1600 cm^{-1} to 2650 cm^{-1} in Fig. 1.5. Thus the film thickness, b, is found as 16 μm from Eq. (1).

1.3.3.3 SPECULAR REFLECTANCE SPECTRA OF COMMERCIAL AL COATED POLYPROPYLENE FILM 2

The specular reflectance spectra of the surfaces of the commercial film 2 are seen in Fig. 1.8; curve 1 show the specular reflectance spectrum of the polypropylene side of the film and curve 2 show the specular reflection spectrum of the Al side of the film. While the polypropylene surface has the spectrum of polypropylene, the Al coated surface reflected the infrared rays. The absorbance values close to zero indicated that the light was not absorbed but reflected by the Al surface.

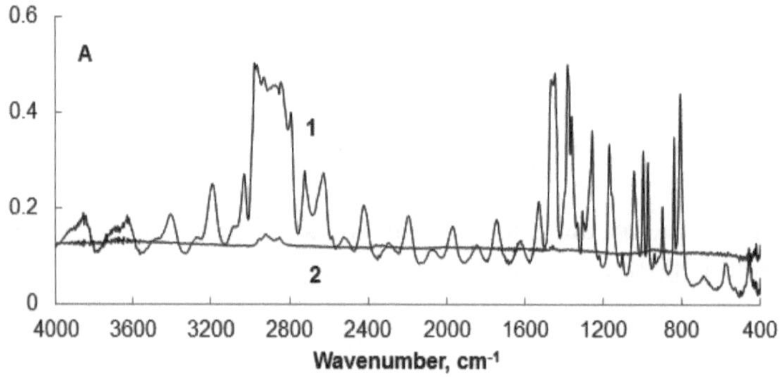

FIGURE 1.8 Specular reflectance spectra of the film 2 (1) polypropylene surface (2) Al coated surface of the film 1.

1.3.3.4 THICKNESS OF COMMERCIAL FILM 2

There were six refraction fringes observed in specular reflection spectrum of commercial Film 2 in Fig. 1.5. Using Equation 1 and inserting values of six fringes between 1600 cm^{-1} to 2650 $cm^{-1}$and the film thickness was calculated as 19 μm.

1.3.3.5 SPECULAR REFLECTANCE SPECTRA OF MAGNETRON SPUTTERED FILMS

The specular reflection spectra of uncoated and coated surfaces of polypropylene film prepared by Ozmihci et al. [5] are seen in Fig. 1.9. Both surfaces showed the characteristic spectrum of polypropylene. This indicated that there were uncoated polypropylene regions on the coated surface. However, the Al coated surface had lower absorbance values at all wave numbers due to reflection of infrared rays from its surface. However, the reflection extent was not as high as the reflection extent of the commercial films. The Al surface of this film was not a reflecting surface like commercial films.

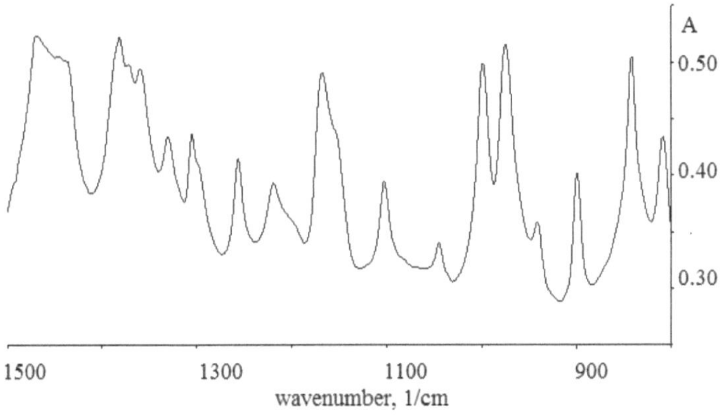

(a)

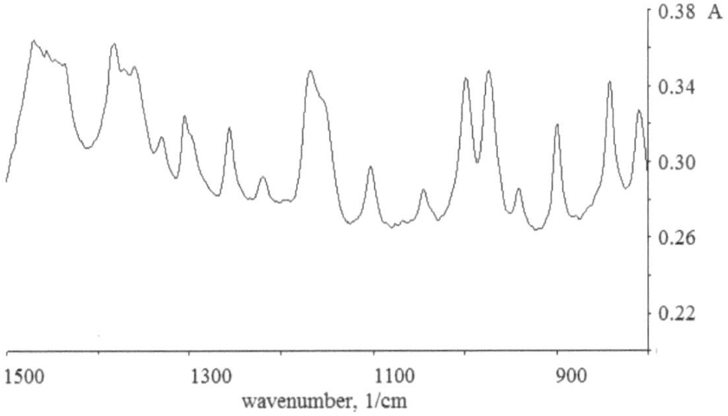

(b)

FIGURE 1.9 Specular reflection spectra of (a) uncoated surface (b) Al coated surface of cast film.

1.3.3.6 SPECULAR REFLECTANCE SPECTRA OF MILK PACKAGE

Milk cardboard container has an Al lid to open it. The specular reflection spectra of uncoated and coated surfaces of milk package are seen in Fig. 1.10. Al coating at the upper surface has lower absorbance value than polymer at the lower surface. Thus Al coating reflected the infrared light and caused light protection of milk from heat and light conditions. The lower surface is less bright. Both surfaces appear to be covered with different polymer layers. The shining surface had peaks at 2920 cm^{-1}, 2860 cm^{-1}, 1745 cm^{-1}, 1645 cm^{-1}, 1282 cm^{-1}, 1070 cm^{-1}, 997 cm^{-1}, 869 cm^{-1}. There are CH_2 stretching vibration peaks at 2920 and 2860 cm^{-1}, C=O peaks at 1745 cm^{-1}, C=C peak at 1645 cm^{-1}. The other surface had peaks at 2951 cm^{-1}, 2885 cm^{-1}, 1734 cm^{-1}, 1454 cm^{-1}, 1296 cm^{-1} and 723 cm^{-1}. 2951 cm^{-1}, 2885 cm^{-1} peak belonged to CH_2 stretching vibrations and 1745 cm^{-1} peak belonged to C=O stretching vibration. Further characterizations are needed to determine which polymers were coated on the surfaces.

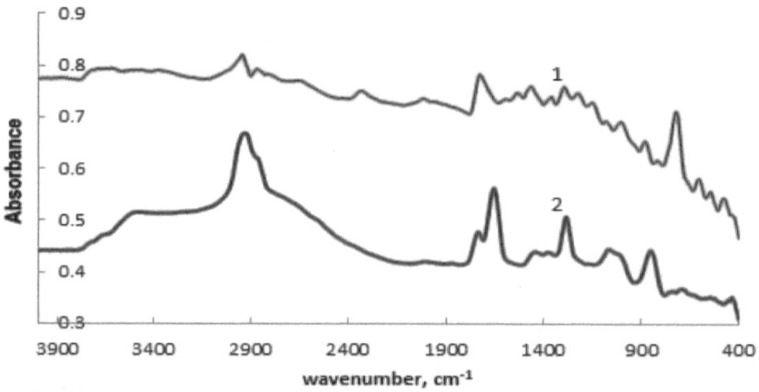

FIGURE 1.10 Specular reflection of milk package 1, Mat surface 2, Shining surface.

1.3.3.7 SPECULAR REFLECTANCE SPECTRA OF CHOCOLATE PACKING

The chocolate packing showed that Al coated surface had lower absorbance values than uncoated surface indicating that infrared light was reflected by the Al coating as seen in Fig. 1.11.

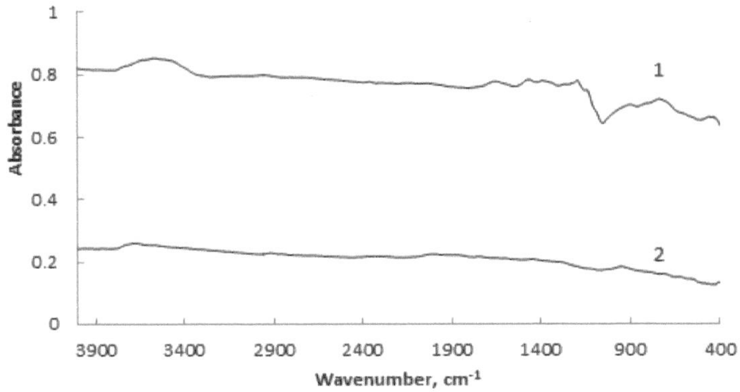

FIGURE 1.11 Specular Reflection Spectra of chocolate package (1) Polymer surface (2) Al coated surface.

1.3.4 COMPARISON OF ABSORBANCE VALUES IN SPECULAR REFLECTANCE SPECTRA OF PACKING MATERIALS

Comparisons of five different films absorbance values are tabulated in Table 1.1. Base line absorbance values at 920 cm^{-1} and maximum absorbance values due to polypropylene band at 1450 cm^{-1} are reported in the table. The base line absorbance values of the uncoated and coated surfaces at 920 cm^{-1} are close to each other for all five films. The highest base line value was observed for magnetron sputtered films. The lowest base line absorbance was observed for commercial film 1. The absorbance values of the uncoated polypropylene surface of the five films at 1450 cm^{-1} are close to each other in the range of 0.50–0.77. This peak is due to bending vibration of the CH_2 groups in polypropylene phase. FTIR light was reflected by the Al surfaces of the films. The highest reflection and the lowest absorbance value of −0.08 werè observed for commercial film 2. Commercial film 1 also had a low absorbance value of 0.11 and high level of reflection. The magnetron sputtered films reflection was not as high as the commercial films, since its absorbance value at 1450 cm^{-1} was 0.36. The Al coating on this film did not cover the whole surface and there were polypropylene phase exposed to infrared light. Chocolate packing materials Al coated surface also had low absorbance value, 0.20 indicating that it also reflected strongly the infrared light. The milk cover Al coated surface even higher absorbance value 0.44 than other films.

TABLE 1.1 Absorbance Values of Specular Reflectance Spectra of Uncoated and Al Coated Surfaces of Packing Materials

Type of film	Absorbance at 1450 cm^{-1}		Absorbance at 920 cm^{-1}	
	Uncoated surface	Al Coated Surface	Uncoated Surface	Al Coated Surface
Commercial film 1	0.50	0.11	0.12	0.11
Commercial film 2	0.66	–0.08	–0.03	–0.07
Magnetron sputtered film	0.60	0.36	0.29	0.28
Chocolate packing	0.77	0.20	0.69	0.18
Milk Cover	0.75	0.44	0.63	0.38

1.4 CONCLUSIONS

Infrared spectroscopy is an efficient tool for measuring the thickness of thin polymer films and their ability to absorb or reflect infrared lights. The thicknesses of the commercial films coated with aluminum were determined to be 16 and 19 μm for film 1 and 2. The aluminum-coated surface of the commercial films had the ability of reflecting the infrared rays, which strike. They can be used efficiently as infrared light shields for the materials inside their packing. The Al coatings obtained by chemical vapor deposition were 32 nm and 185 nm for commercial film 1 and 2, respectively. The coatings were more perfect than the coating obtained by magnetron sputtered cast film. The Al coating of the cast film was more brittle, and there were uncoated polypropylene regions and the infrared light was only partially filtered. Thus it was a less efficient infrared shield compared to commercial films. The magnetron sputtering method needs more investigation for optimum results. The chocolate and milk packing materials Al coated surfaces reflected the infrared light better than their other surfaces. They were also good infrared light protectors. However, milk-packing material Al surface was also coated with another polymer layer either for esthetic or safety reasons or since it was used in contact with milk.

ACKNOWLEDGMENTS

The authors thank The POLİNAS and POLİBAK companies and the authors of reference 19 for providing for providing commercial aluminum coated films and magnetron sputtered film, respectively.

KEYWORDS

- **Absorption**
- **Commercial films**
- **Magnetron films**
- **Polymer films techniques**
- **Reflection of infrared light**
- **Uncoated and al coated surfaces**

REFERENCES

1. Aihong, G., Xuejiao T., & Sujuan, Z. (2011). *Key Engineering Materials*, *474–476*, 195–199.
2. Dombrowsky, L. A. (1998). *Rev. Gen. Therm*, *37*, 925–933.
3. King, D. E., Drewry, D. G., Sample, J. L., Clemons, D. E., Caruso, K. S., Potocki, K. A., Eng, D. A., Mehoke, D. S., Mattix, M. P., Thomas, M. E., & Nagle, D. C. (2009). *Int. J. Appl. Ceramic Tech, 6(3),* 355–361.
4. Shahidi, S., Ghoranneviss, M., Moazzenchi, B., Anvari, E., & Rashidi, A. (2007). *Surface and Coating Technology*, *201*, 5646–5650.
5. Takano, I., Inoue, N., Matsui, K., Kokubu, S., Sasase, M., & Isobe, S. (1994). *Surface and Coating Technology*, *66,* 509–513.
6. O'Hare, L., A., Leadley, S., & Parbhoo, B. (2002). *Surface Interface Anal*, *33*, 335–342.
7. Greer, J., & Street, R. A. (2007). *Acta Materialia, 55*, 6345–6349
8. Fortunato, E., Nunes, P., Marques, A., Costa, D., Aguas, H., Ferreira, I., Costa, M. E. V., Godinho, M., H., Almeida, P. L., Borges, J. P., & Martins, R. (2002). *Surface and Coating Technology*, *151–152*, 247–251.
9. Yanaka, M., Henry, B. M., Roberts, A. P., Grovenor, C. R. M., Briggs, G. A. D., Sutton, A. P., Miyamoto, T., Tsukahara, Y., Takeda, N., & Chjater, R. J. (2001). *Thin Solid Films*, 397, 176–185.
10. Bichler, C. H., Kerbstadt, T., Langowski, H. C., & Moosheimer, U. (1999). *Surface and Coating Technology*, 112, 373–378.

11. Bichler, C. H., Langowski, H. C., Moosheimer, U., & Seifert, B. (1997). *Journal of Adhesion Science and Technology, 11(2)*, 233–246.
12. Moosheimer, U., & Bichler, C. H. (1999). *Surface and Coating Technology, 116*, 812–819.
13. Oishi, T., Goto, M., Pihosh, Y., Kasahara, A., & Tosa, M. (2005). *Applied Polymer Science, 241*, 223–226.
14. Qi-Jia He., Ai-Min Zhang, & Ling-Hong Guo. (2004). *Polymer–Plastics Technology and Engineering, 43(3)*, 951–961.
15. Qui, J., Liu, L. H., & Hsu, P. F. (2005). *Appl. Surf. Sci., 111*, 1912–1920.
16. Rahmatollahpur, R., Tahidi, T., & Jamshidi-Chaleh, K. (2010). *J. Mat. Sci., 45*, 1937.
17. Ozmihci, F., Balkose, D., & Ulku, S. (2001). *J. Appl. Polym. Sci., 82*, 2913–2921.
18. Balkose, D., Oguz, K., Ozyuzer, L., Tari, S., Arkis, E., & Omurlu, F. O. (2012). *J. Appl. Polym. Sci., 120*, 1671–1678.
19. http://www.piketech.com/skin/Fashion_Mosaic_Blue/Application-pdfs/Calculating Thickness-Free Standing Films-by FTIR.pdf, 2012.
20. Luongo, J. P. (1960). Infrared Study of Polypropylene. *J. App. Polym. Sci., 9*, 302–309.

CHAPTER 2

SPECIFIC FEATURES OF NOVEL BLENDS ON THE BASIS OF COPOLYMERS

SVETLANA G. KARPOVA, ALEKSEI A. IORDANSKII,
SERGEI M. LOMAKIN, and ANATOLII A. POPOV

CONTENTS

ABSTRACT

The specific features of novel blends on the basis of copolymers 3-hydroxybutyrate with hydroxyl valerate (95:5 mol %) (PHBV) and segmented polyetherurethane (SPEU) have been investigated. The samples with different content (100/0, 60/40, 40/60, 50/50, and 0/100 wt.%) are explored by structural (WAXS, FTIR), thermo physical (DSC) and dynamic (probe ESR nano analysis, H–D isotopic exchange) methods at about physiological (40 °C) and elevated (70 °C) temperatures after treatment by water or heavy water. It was shown that for the first 5 h during water exposition at 40 °C the individual polymers (PHB and SPEU) and the blends kept structural stability that was confirmed by ESR time correlation data for the probe TEMPO and stable crystallinity (WAXS, DSC). Dependence of H–D exchange rate on the blend content has minimum at PHB/SPEU ratio 60/40. At 70 °C the water–temperature effect leads to enhanced molecular mobility of TEMPO in all systems besides the individual PHB. In accordance with DSC data in this case the crystallinity and melting points of PHBV are decreased. The descriptions of PHBV–SPEU system using a combination of structural and dynamic features gives a more complete assessment of its structural evolution at relatively short-term times, which precede its hydrolytic decomposition.

2.1 INTRODUCTION

The new biodegradable blends based on the combination of synthetic and natural polymers are alternative option for the use of individual polymers. As a result of composite production it is expected the emergence of essentially new exploitation characteristics, which are not inherent to the original polymer components. Biodegradable systems are widely used in innovative technologies for the drug delivery, in tissue engineering, and prosthetic vascular stenting, in a contemporary design of environmentally friendly barrier materials in packaging [1–3]. When creating such systems, segmented polyetherurethanes (conventional SPEU on the basis of multifunctional isocyanates) may be of particular interest. Because of the unique combination between mechanical properties and biocompatibility, they are widely and successfully applied in various fields of biomedicine as constructional and functional materials [4, 5]. However, the low rate of SPEU biodegradation, which is a positive factor in the case of a long-term functioning, limits considerably their use during a short-term exploitation. Regulation of SPEU lifetime can be achieved by blending it with a variety of biodegradable biopolymers, such as poly(α-hydroxyacid)s or poly(β-hydroxyalkanoate)s and the common representatives of the latter as poly(3-hydroxybutyrate) (PHB) and its copolymer with 3-hydroxyvalerate (PHBV) [6, 7].

PHB, along with valuable properties, has certain disadvantages, in particular, high cost and fragility. For overcoming such disadvantages either copolymers PHAs or their blends with other natural polymers, in particular, with chitosan [8] are often used.

PHB–SPEU and PHBV–SPEU polymer systems can be applied as the basis of promising new composites, in particular, for use in cardiosurgery, as well as for a scaffold design. By varying the blend content, and, consequently, affecting the morphology of the systems, we can obtain composition with different physical and chemical characteristics such as permeability, solubility in water and drugs, controlled release, reduced rate of degradation. The comprehensive description of PHBV-SPEU system requires the assessment of a combination of structural and dynamic features that allow a more complete evaluation of its structural evolution at relatively short-term times that precede its hydrolytic decomposition as a result of exposure to water at elevated temperatures.

Blends comprising the copolymer of 3-hydroxybutyrate (~95 mol%) and 3-hydroxyvalerat (~ 5 mol%) – PHBV (Tianan Co, Ningbo, China), $Mw = 2.4\times105$, $Mn = 1.5\times105$, $\rho = 1.25$ g/cm^3) and SPEU on the basis of MDI medical grade (Elastogran, Basf Co., Germany), $Mw = 2.29\times105$, $Mn = 5.3\times104$, $\rho = 0.97$ g/cm^3 were studied. Mixing weight ratio of PHBV/SPEU ranged in the following sequence: 100:0, 60:40, 50:50, 40:60, 0:100%. The films were obtained in a solvent mixture of tetra-hydrofuran and dioxane. Additionally, by the DSC method the thermo physical characteristics (melting points and endothermal fusion) were studied. DSC measurements were performed with a micro calorimeter DSC 204 F1 Netzsch Co., in an inert atmosphere of argon at a heating rate of 10°/min. Analysis of the mixtures was carried out by dynamic (ESR probe analysis and isotope deuterium D–H exchange) and structural (WAXS and DSC) methods. Segmental mobility was studied by paramagnetic probe by determining the correlation time τ, characterizing the rotational mobility of the nitroxyl probe TEMPO in accordance with known method [9, 10].

X-ray analysis of the films was carried out using transmission technique with a diffractometer Bruker Advance D8 (Cu Kα). IR spectra of deuterated films were recorded on a spectrophotometer Bruker IFS 48 with a Fourier transform at a resolution of 2 cm^{-1}.

2.2 X-RAY AND DSC STUDIES

In this chapter it was shown by X-ray technique (WAXS) that the original samples of PHBV are characterized by high crystallinity. In their diffraction patterns at least 5 reflections of the orthorhombic lattice with characteristics a = 5.74 Å, b = 13.24 Å and c = 5.98 Å were determined that well conform with the earlier studies [11, 12]. In Fig. 2.1, the WAXS diffractograms are presented as asset of curves belonging to initial PHBV (*1*), PHBV treated by water at 40 °C (*2*), PHBV/SPEU blend with component ratio 40/60 (*3*), the same blend treated by water at 70 °C (*4*), and the PHBV/SPEU (60/40) blend also treated by water at 70 °C (*5*). The SPEU presence in the system leads to an amorphous hallo in the range 22° (cf. curves *1* and *3*), its intensity is increased with the SPEU content.

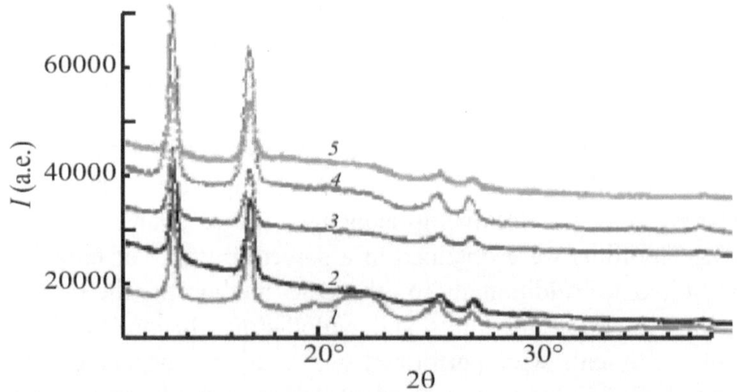

FIGURE 2.1 WAXS diffractograms for initial PHB (*1*), PHB after water treatment at 318 K (*2*), initial blend PHB-SPEU 40:60 (*3*), blend PHB-SPEU 40:60 after water treatment at 343 K (*4*), and PHB-SPEU blend 60:40 after water treatment at 343 K.

The impact of the water medium causes noticeable changes in the crystal structure, namely an increase in the crystallinity degree, the crystallite size growth and improving their crystalline structure that manifested as a de-crease in the basic interplanar spacings *d*020 and *d*110 (Table 2.1). The ratio of the integrated intensity of the crystal-line reflections to the total intensity is 66% for the initial samples and 84% for samples treated with water. Crystallite sizes, the corresponding lattice parameters, and crystallinity of PHBV are presented in Table to summaries the results of X-ray analysis. The above influence of water is similar for hydrophobic PELD with different molecular weights, which were subjected to ionizing radiation [13].

TABLE 2.1 Crystalline and Thermo Physical Characterization of PHBV and Its Blend

PHB content (%)	Crystallinity (%) WAXS-DSC	Interplanar spacings (A)	Sizes of crystallites
100 (initial)	66–58	6,640	215/160
100 (water exposition)	84–59	6,606	255/190
60 (water exposition)	68–58	6,649	183/157
40 (initial)	55–58	6,600	174/149
40 (water exposition)	53–50	6,600	173/145

The exposure of the samples in aquatic environment affects the original PHBV and its composition with SPEU in different ways. The parent bio-copolymer shows stability and even improving of the crystalline phase. According to the DSC data, its crystallinity degree is kept constant after the contact with water and approximately is equal to 60%. After the water treatment for 4 h and elevated heating (70 °C) simultaneously, melting temperature (*T*m) shifts to higher temperatures from 175 to 177 °C for the native sample and from 172 to 174 °C for the sample subjected to the first melting-cooling cycle. These T-scale shifts as well as the analysis by WAXS indicated more perfect organization of PHBV crystals that occurred due to water impact.

For polymer compositions PHBV/SPEU exposed to water at 70 °C, the DSC curves of melting show a moderate decrease in crystallinity from 59 to 50%. The crystallinity reduction is accompanied by a decrease of *T*m from 174 °C for the initial sample to 170 °C for the PHBV/SPEU blend with 60/40% ratio and then to 168 °C for the blend with 40/60% content. Therefore, the SPEU molecules prevent the crystallization completion of PHBV and, hence, reduce both the quality of structural organization and the crystallinity degree.

2.3 DYNAMICS OF ESR PROBE MOBILITY AND H–D EXCHANGE KINETICS IN D_2O

Recently we have studied the behavior of spin probes (TEMPO and TEMPOL) in high crystalline PHB at room and elevated temperatures [14]. Water–temperature effects on molecular mobility of the ESR probe (TEMPO) in PHBV were studied at two different temperatures: 40 °C, close to physiological temperature, and 70 °C adopted in a number of studies as the standard temperature for accelerated testing of hydrolytic stability of biopolymers [15, 16]. Figure 2.2 shows that the heating of polymer compositions in water for 4 h at ~40 °C does not significantly reduce the molecular mobility of the probe in the mixture, as well as does not change the PHBV crystallinity that we have shown above by DSC technique. For samples of PHBV and its blends, the impact of the aquatic environment under more rigorous conditions, that is, at 70 °C, is accompanied by the probe mobility increase in comparison with the mobility in the initial polymer systems. These results obtained from ESR spectra reveal plasticizing effect of water molecules, so that segmental mobility of macromolecules in the intercrystalline space is increased and hence the relaxation processes proceeds faster that lead to an increase in the rotation velocity of the probe TEMPO.

The supplemental feature of SPEU behavior in the blends is a FTIR band intensity decrease for the –NH– fragments of urethane groups, which can be represented as a series of kinetic curves. The curves reflect the H–D exchange rate in the PHB/SPEU blends immersed into heavy water. Only those –NH– groups can exchange proton for deuterium that are accessible to the attack by molecules of D_2O [17].

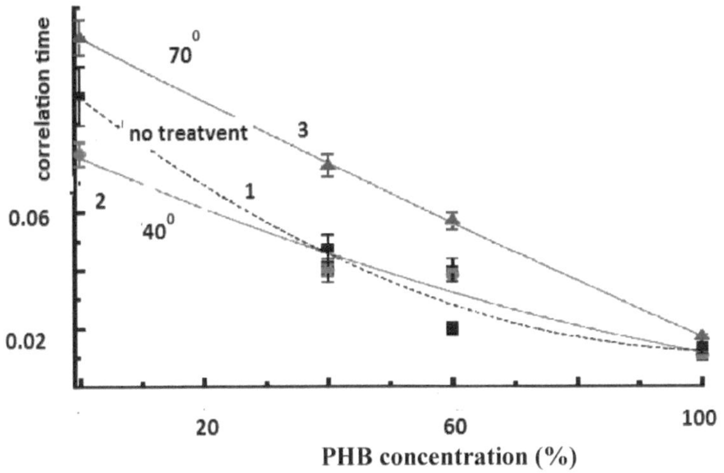

FIGURE 2.2 ESP correlation frequency of TEMPO in the parent polymers (PHB, SPEU) and their blends (1 – initial polymer systems, 2 – water treatment at 40°C, 3 – water treatment at 70°C).

These fragments belong to the amorphous regions and are not included in the domain structures of SPEU. For the blend films the dependence of exchange rate on the PHBV/SPEU composition has an extreme character (Fig. 2.3), and the position of the minimum is in the same concentration range as the minimum value of exchange degree (not shown).

It was at this ratio of PHBV/SPEU (40:60) that the most ordered structure in the blend was formed and stabilized by intermolecular hydrogen bonds. Figure 2.3 also shows that in this concentration range there is a maximum oscillation frequency shift ascribed to –NH– fragments in relation to the frequency of –NH– groups in the native SPEU. Such a change in the dynamics of fluctuations is probably due to the formation of the previously mentioned hydrogen bonds in the system.

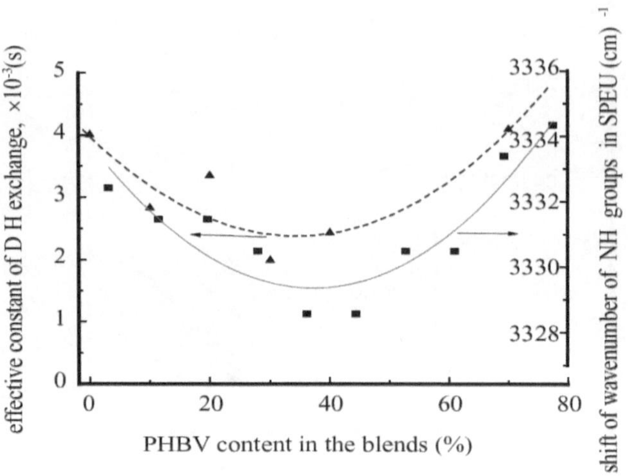

FIGUER 2.3 Dependences of effective constant for isotopic exchange (left *y*-axis) and the shift of wave number for –NH–fragment in urethane groups (right *y*-axis) on PHBV concentration in PHBVSPEU blends.

2.4 CONCLUSIONS

The complex dynamic and structural properties reveal the influence of SPEU on the molecular dynamics and structure of PHBV, by forming intermolecular hydrogen bonds and setting up of energy and steric barriers to PHBV crystallization. In terms of design of new biomedical composites, results presented have a scientific and practical interest to describe their behavior at water–temperature exposure.

KEYWORDS

- **Biodegradable composition**
- **Crystallinity**
- **DSC**
- **EPR**
- **X-ray analysis**

REFERENCES

1. Tian, H., Tang, Z., et al. (2012). *Progress in Polymer Science*, *37*, 237. doi:10.1016/j.progpolymsci.2011.06.004.
2. Suyatma, N. E., Copinet, A., et al. (2004). *J. Polym. Environ, 12(1),* 1. doi: 1566–2543/04/0100–01/0.
3. Bonartsev, A. P., Boskhomodgiev, A. P., et al. (2012). *Molecular Crystals and Liquid Crystals, 556(1),* 288
4. Bagdi, K., Molnar, K., et al. (2011). *EXPRESS Polymer Letters, 5(5),* 417 doi: 10.3144/expresspolymlett.2011.41.
5. Shi, R., Chen, D., et al., (2009). *Int. J. Mol. Sci., 10,* 4223; doi: 10.3390/ijms10104223.
6. Sanche, M. D-Garcia, Gimene, E., et al. (2008). *Carbohydrate Polymers, 71(2),* 235 doi: 10.1016/j.carbpol. 2007.05.041.
7. Yang, K-K., Wang, X.-L., et al. (2007). *J. Ind. Eng. Chem, 13(4),* 485
8. Iordanski, A. L., Rogovina, S. Z., et al. (2010). *Doklady Physical Chemistry, 431,* Part 2, 60. doi: 10.1134/S0012501610040020.
9. Smirnov, A. I., Belford, R. L., et al. (1998). (Berliner, L. J., Ed.). Plenum Press, New York, Ch. 3, 83–108.
10. Karpova, S. G., Iordanskii, A. L., et al. (2012). *Russian J. Physical Chemistry B, 6(1),* 72.
11. Bloembergen, S., Holden, D. A., et al. (1989). Marchessault: *Macromolecules*, 22, 1656
12. Di Lorenzo, M. L. Raimo, M., et al. (2011). *J. Macromol. Sci. Phys, B40(5),* 639.
13. Selikhova, V. I., Shcherbina, M. A., et al. (2002). *Polymer Chem, 22(4),* 605 (in Russian).
14. Kamaev, P. P., Aliev, I. I., et al. (2001). *Polymer, 42,* 515.
15. Freier, T., Kunze, C., et al. (2002). *Biomaterials, 23(13),* 2649
16. Artsis, M. I., Bonartsev, A. P., et al. (2012). *Mol. Cryst. Liq. Cryst, 555,* 232 DOI: 10.1080/15421406.2012.635549.
17. Zaikov, G. E., Iordanskii, A. L., et al. (1988). *Diffusion of Electrolytes in Polymers.* Ser. New Concepts in Polymer Science. Utrecht-Tokyo, VSP BV, 229–231.

CHAPTER 3

INTERRELATION BETWEEN THE PARTICLE SIZE OF A TITANIUM CATALYST AND ITS KINETIC HETEROGENEITY IN THE POLYMERIZATION OF ISOPRENE

ELENA M. ZAKHAROVA, VADIM Z. MINGALEEV, and
VADIM P. ZAKHAROV

CONTENTS

ABSTRACT

The effect particle size of micro heterogeneous catalyst $TiCl_4$–Al(iso-$C_4H_9)_3$ on the basic patterns of isoprene polymerization is studied. Fraction of particles with a certain size isolated from the mixture of particles, which is formed by reacting the initial components of the catalyst and its following exposition. The most active in isoprene polymerization is particles fraction with starting diameter of 0.7–4.5 μm. In these particles preferably localized highly active polymerization site. Doping diphenyloxide and piperylene, lower exposition temperature, hydrodynamic actions in turbulent flows result in formation of nearly mono disperse catalyst with diameter of 0.15–0.18 μm. In this case there is a shift in activity spectrum of polymerization sites towards formation of single site with high activity.

PACS: 82.65.+r, 82.35.-*x*.

3.1 INTRODUCTION

The formation of highly stereo regular polymers under the action of micro heterogeneous Ziegler–Natta catalysts is accompanied by broadening of the polymer MWD [1, 2]. This phenomenon is related to the kinetic heterogeneity of active sites (AC) [1, 3, 4]. The possible existence of several kinetically nonequivalent AC of polymerization correlates with the nonuniform particle size distribution of a catalyst [4]. At present time much attention is given to study the influence of micro heterogeneous catalysts particle size on the properties of polymers [5, 6]. However, almost no detailed study of the effect particle size catalysts on their kinetic heterogeneity.

The micro heterogeneous catalytic system based on $TiCl_4$ and Al (iso-$C_4H_9)_3$ that are widely used for the production of the cis-1,4-isoprene. The research [7] show that the targeted change of the solid phase particle size during the use of a tubular turbulent reactor at the stage of catalyst exposure for many hours is an effective method for controlling the polymerization process and some polymer characteristics of isoprene. We suppose that the key factor is the interrelation between the reactivity of isoprene polymerization site and the size of catalyst particles on which they localize.

The aim of this chapter was to investigate the interrelation between the particle size of a titanium catalyst and its kinetic heterogeneity in the polymerization of isoprene.

3.2 EXPERIMENTAL PART

Titanium catalytic systems (Table 3.1) were prepared through two methods.

Method 1

At 0 or –10°C in a sealed reactor 30–50 mL in volume with a calculated content of toluene, calculated amounts of $TiCl_4$ and Al(*iso*-$C_4H_9)_3$ toluene solutions (cooled to the same temperature) were mixed. The molar ratio of the components of the catalyst corresponded to its maximum activity in isoprene polymerization. The resulting catalyst was kept at a given temperature (Table 3.1) for 30 min under constant stirring.

TABLE 3.1 Titanium Catalytic Systems and Their Fractions used for Isoprene Polymerization

Catalyst	Labels	Molar ratio of catalyst components			T, °C	Method	Range of particle diameters in fractions of titanium catalysts, μm		
		Al/Ti	DPO/Ti	PP/Al			Fraction I	Fraction II	Fraction III
$TiCl_4$–Al(i-$C_4H_9)_3$	C-1	1	-	-	0	1	0.7–4.5	0.15–0.65	0.03–0.12
						2	–	0.20–0.7	0.03–0.18
$TiCl_4$–Al(i-$C_4H_9)_3$–DPO	C-2	1	0.15	-	0	1	0.7–4.5	0.15–0.65	0.03–0.12
						2	–	0.15–0.68	0.03–0.12
$TiCl_4$–Al(i-$C_4H_9)_3$–DPO–PP	C-3	1	0.15	0.15	0	1	–	0.12–0.85	0.03–0.10
						2	–	0.15–0.80	0.03–0.12
$TiCl_4$–Al(i-$C_4H_9)_3$–DPO–PP	C-4	1	0.15	0.15	-10	1	–	0.12–0.45	0.03–0.10
						2	–	0.12–0.18	0.04–0.11
Averaged ranges, μm							0.7–4.5	0.15–0.69	0.03–0.14

Note: DPO – diphenyloxide, PP – piperylene

Method 2

After preparation and exposure of titanium catalysts via method 1, the system was subjected to a hydrodynamic action via single circulation with solvent through a six-section tubular turbulent unit of the diffuser-confusor design [8] for 2–3 s.

The catalyst was fractionated through sedimentation in a gravitational field. For this purpose calculated volumes of catalysts prepared through methods 1 and 2 were placed into a sealed cylindrical vessel filled with toluene. In the course of sedimentation, the samples were taken from the suspension column at different heights, a procedure that allowed the separation of fractions varying in particle size.

The titanium concentrations in the catalyst fractions were determined via FEK colorimeter with a blue light filter in a cell with a 50 mm thick absorbing layer. A K_2TiF_6 solution containing 1×10^{-4} g Ti/mL was used as a standard.

The catalyst particle size distribution was measured via the method of laser diffraction on a Shimadzu Sald-7101 instrument.

Before polymerization, isoprene was distilled under a flow of argon in the presence of Al(*iso*-$C_4H_9)_3$ and then distilled over a $TiCl_4$–Al(*iso*-$C_4H_9)_3$ catalytic system, which provided a monomer conversion of 5–7%. The polymerization on fractions of the titanium catalyst was conducted in toluene at 25 °C under constant stirring. In this case, the calculated amounts of solvent, monomer, and catalyst were successively placed into a sealed ampoule 10–12 mL in volume. The monomer and catalyst concentrations were 1.5 and 5×10^{-3} mol/L, respectively. The polymerization was terminated via the addition of methanol with 1% ionol and 1% HCl to the reaction mixture. The polymer was repeatedly washed with pure methanol and dried to a constant weight. The yield was estimated gravimetrically.

The MWD of poly isoprene was analyzed via GPC on a Waters GPC-2000 chromatograph equipped with three columns filled with a Waters micro gel (a pore size of 103–106 A) at 80°C with toluene as an eluent. The columns were preliminarily calibrated relative to Waters PS standards with a narrow MWD (M_w/M_n = 1.01). The analyzes were conducted on a chromatograph, which allows calculations with allowance for chromatogram blurring. Hence, the need for additional correction of chromatograms was eliminated.

The microstructure of poly isoprene was determined via high-resolution 1H NMR spectroscopy on a Bruker AM-300 spectrometer (300 MHz).

The MWD of cis-1,4-polyisoprene obtained under the aforementioned experimental conditions, $q_w(M)$, were considered through the equation

$$q_w(M)=)=\int_0^\infty \Psi(\beta)M\beta^2\exp(-M\beta)d\beta$$

where β is the probability of chain termination and ψ(β) is the distribution of active site over kinetic heterogeneity, M is current molecular weight.

As was shown previously [9] Eq. (1) is reduced to the Fredholm integral equation of the first kind, which yields function ψ(β) after solution via the Tikhonov regularization method. This inverse problem was solved on the basis of an algorithm from [9]. As a result, the function of the distribution over kinetic heterogeneity in ψ(lnβ)–lnM coordinates with each maximum related to the functioning of AC of one type was obtained.

3.3 RESULTS

After mixing of the components of the titanium catalyst, depending on its formation conditions, particles 4.5 μm to 30 nm in diameter, which are separated into three arbitrary fractions, are formed (Table 3.1).

During the formation of catalyst C-1 via method 1, the fraction composed of relatively coarse particles, fraction I constitutes up to 85% (Fig. 3.1). In method 2, the hydrodynamic action on the titanium catalyst formed under similar conditions results in an increase in the content of fraction II. Analogous trends are typical of catalyst C-2. The catalyst modification with piperylene additives, catalyst C-3, is accompanied by the disappearance of fraction I and an increase in the content of fraction II (Fig. 3.1), as was found during the hydrodynamic action on C-1. The hydrodynamic action on a two-component catalyst is equivalent to the addition of piperylene to the catalytic system. The preparation of catalytic complex C-3 via method 2 results in narrowing of the particle size distribution of fraction II owing to disintegration of particles 0.50–0.85 μm in diameter (Fig. 3.1). The reduction of the catalyst exposure temperature to −10°C (catalyst C-4) is accompanied by further disintegration of fraction II (Fig. 3.1). In this case the content of particles 0.19–0.50 μm in diameter decreases to 22% with predominance of particles 0.15–0.18 μm in diameter. The formation of C-4 via method 2 results in additional dispersion of particles of fraction II, with the content of particles 0.15–0.18 μm in diameter attaining 95%.

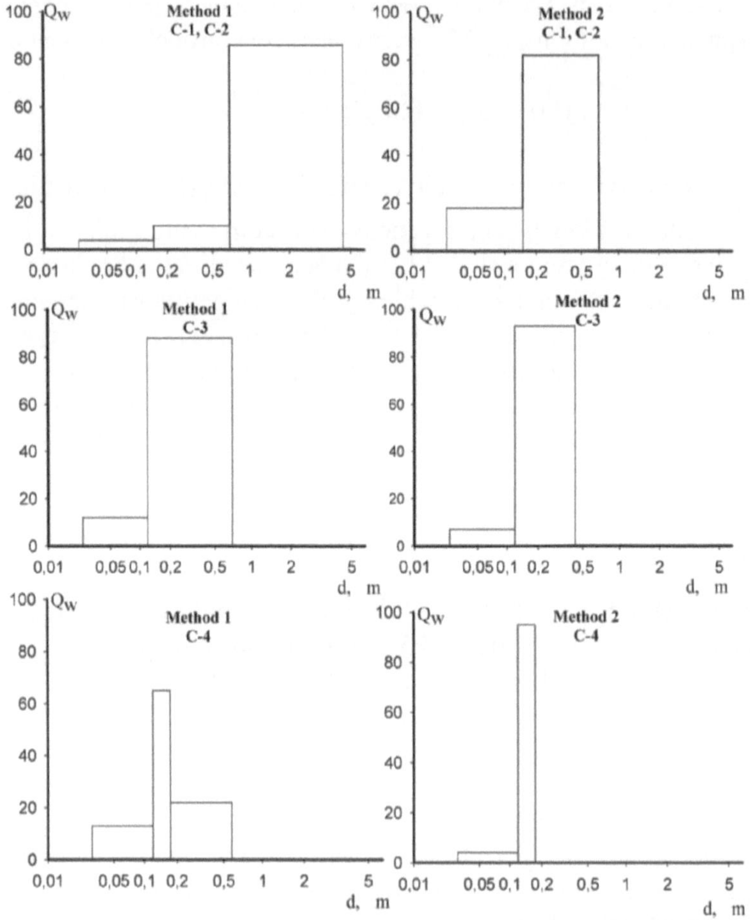

FIGURE 3.1 Fractional compositions of titanium catalyst C-1–C-2 (Table 3.1).

The content of the finest catalyst particles in the range 0.03–0.14μm, fraction III, attains 5–12% and is practically independent of the catalyst formation conditions. The most considerable changes are shown by particles 0.18–4.50 μm in diameter. Particles of fraction I are easily dispersed as a result of the hydrodynamic action in turbulent flows and the addition of catalytic amounts of piperylene, and their diameter becomes equal to that of particles from fraction II. The decrease in the catalyst exposition temperature from 0 to 10°C with subsequent hydrodynamic action leads to a more significant reduction of particle size and the formation of a narrow fraction.

Isolated catalyst fractions differing in particle size were used for isoprene polymerization. The cis-1,4-polymer was obtained for all fractions, regardless of their formation conditions. The contents of cis-1,4 and 3,4 units were 96–97 and 3–4%, respectively. Coarse particles (fraction I) are most active in isoprene polymerization (**method 1**) on different fractions of C-1 (Fig. 3.2).

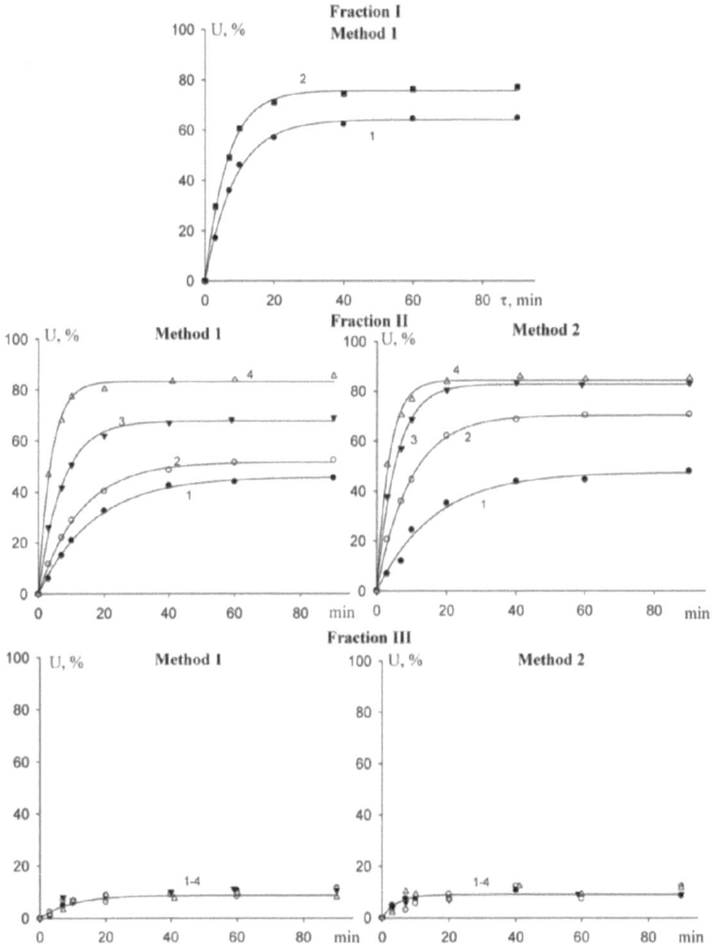

FIGURE 3.2 Cis-1,4-Polyisoprene yields U vs. polymerization times τ in the presence of fractions particles of titanium catalysts (*1*) C-1, (*2*) C-2, (*3*) C-3, and (*4*) C-4.

As the particle size of C-1 decreases, its activity drops significantly. The catalyst modification with diphenyloxide (C-2) has practically no effect on the fractional composition, but the activities of different catalyst fractions change. The most marked increase in activity was observed for fraction I. Catalyst C-3 prepared via method 1 comprises two fractions, with fraction II having the maximum activity. The decrease of the catalyst exposition temperature to –10°C (C-4) result in further increase in the rate of isoprene polymerization on particles of fraction II.

The hydrodynamic action increases the content of fraction II in C-1, but its activity in isoprene polymerization does not increase (Fig. 3.2). For C-2 the analogous change in the fractional composition is accompanied by an increased activity of fraction II (**method 2**). The addition of piperylene to C-3 results in a stronger effect on the activity of fraction II under the hydrodynamic action. The change of the hydrodynamic regime in the reaction zone does not affect the activity of fine particles of catalyst fraction III. Isoprene polymerization in the presence of fraction III always has a low rate and a cis-1,4-polyisoprene yield not exceeding 7–12%.

During polymerization with C-1 composed of fraction I (method 1), the weight-average molecular mass of polyisoprene increases with the process time (Table 3.2). Polyisoprene prepared with catalyst fraction II has a lower molecular mass. A more considerable decrease in the weight-average molecular mass is observed during isoprene polymerization on the finest particles of fraction III, with M_w being independent of the polymerization time. The width of the MWD for polyisoprene obtained on fraction I increases with polymerization time from 2.5 to 5.1. The polymer prepared in the presence of catalyst fraction III shows a narrow molecular mass distribution ($M_w/M_n \sim 3.5$). The addition of DPO is accompanied by an increase in the weight-average molecular mass of polyisoprene obtained in the presence of fraction I. Under a hydrodynamic action (method 2) on catalysts C-1 and C-2, polyisoprene with an increased weight-average molecular mass is formed. During the addition of piperylene to the catalyst (C-3), M_w of polyisoprene increases to values characteristic of C-2 formed via method 2. The formation of the catalytic system $TiCl_4$–Al(*iso*-$C_4H_9)_3$–DPO–piperylene via method 1 at −10°C results in substantial increases in the activity of the catalyst and the reactivity of the active centers of fraction II. The polymerization of isoprene on catalyst fraction III, regardless of the conditions of catalytic system formation (electron donor additives, exposure temperature, hydrodynamic actions), yields a low molecular mass polymer with a polydispersity of 3.0–3.5.

TABLE 3.2 Molecular-Mass Characteristics of *Cis*-1,4-Polyisoprene 1 and 2 Are the Methods of Catalyst Preparation

C	τ min	Fraction I				Fraction II				Fraction III			
		$M_w \times 10^{-4}$		M_w/M_n		$M_w \times 10^{-4}$		M_w/M_n		$M_w \times 10^{-4}$		M_w/M_n	
		1	*2*	*1*	*2*	*1*	*2*	*1*	*2*	*1*	*2*	*1*	*2*
C-1	3	17.1		2.6		43.2	43.6	4.6	4.8	25.4	21.4	3.7	3.8
	20	56.3		4.2		46.5	57.5	4.0	5.9	23.7	22.6	3.4	3.7
	40	64.9		4.6		47.4	62.7	4.1	5.8	27.6	28.9	3.6	3.6
	90	72.2		5.1		54.3	61.4	4.3	6.2	26.8	22.4	3.7	3.6
C-2	3	24.3		3.3		39.5	54.8	4.3	4.9	21.1	22.3	3.8	3.7
	20	69.7		3.8		41.2	67.5	3.9	4.6	24.6	24.6	3.7	3.4
	40	78.4		3.7		42.6	72.8	4.4	4.3	26.3	22.4	3.3	3.5
	90	84.6		4.0		58.0	73.2	4.3	4.2	24.7	24.3	3.5	3.3
C-3	3					66.2	60.5	3.9	3.8	22.4	20.0	3.8	3.6
	20					75.4	71.1	3.7	3.5	28.1	21.3	3.7	3.7
	40					79.6	78.8	4.1	4.0	21.5	23.6	3.8	3.8
	90					81.6	79.6	3.9	3.8	23.2	26.8	3.6	3.3
C-4	3					81.2	72.7	3.6	3.7	25.4	20.3	3.8	3.6
	20					62.7	51.3	3.3	3.2	24.8	21.6	3.7	3.7
	40					57.6	44.3	3.1	3.2	23.2	26.8	3.6	3.6
	90					61.4	48.2	3.2	3.4	19.6	21.5	3.1	3.3

The solution of the inverse problem of the formation of the MWD in the case of cis-1,4-polyisoprene made it possible to obtain curves of the active site distribution over kinetic heterogeneity (Figs. 3.3–3.6). As a result of averaging of the positions of all maxima three types of polymerization site that produce isoprene macromolecules with different molecular masses were found:

type A ($\ln M = 10.7$), type B ($\ln M = 11.6$), and type C ($\ln M = 13.4$).

The polymerization of isoprene in the presence of fractions I and II of C-1 (method 1) occurs on site of types B and C (Fig. 3.3). The polymerization in the presence of fraction III proceeds on active site of type A only. The single site and low activity character of the catalyst composed of particles of fraction III is typical of all the studied catalysts and all the methods of their preparation. Thus, these curves are not shown in subsequent figures. Catalyst C-2 prepared via method 1 likewise features the presence of type B and type C site in isoprene polymerization on fractions I and II (Fig. 3.4). The hydrodynamic action (method 2) on C-1 accompanied by dispersion of particles of fraction I does not change the types of site of polymerization on particles of fraction II (Fig. 3.5). A similar trend is observed during the same action on a DPO-containing titanium catalyst in turbulent flows (Fig. 3.5).

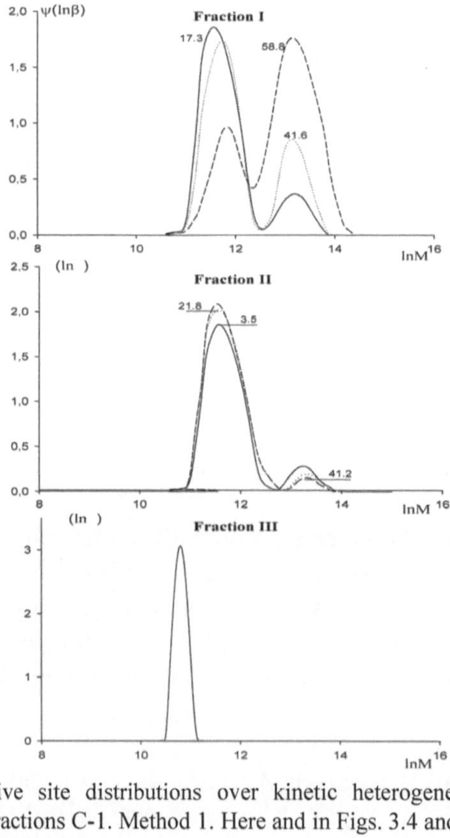

FIGURE 3.3 Active site distributions over kinetic heterogeneity during isoprene polymerization on fractions C-1. Method 1. Here and in Figs. 3.4 and 3.5 numbers next to the curves are conversions (%).

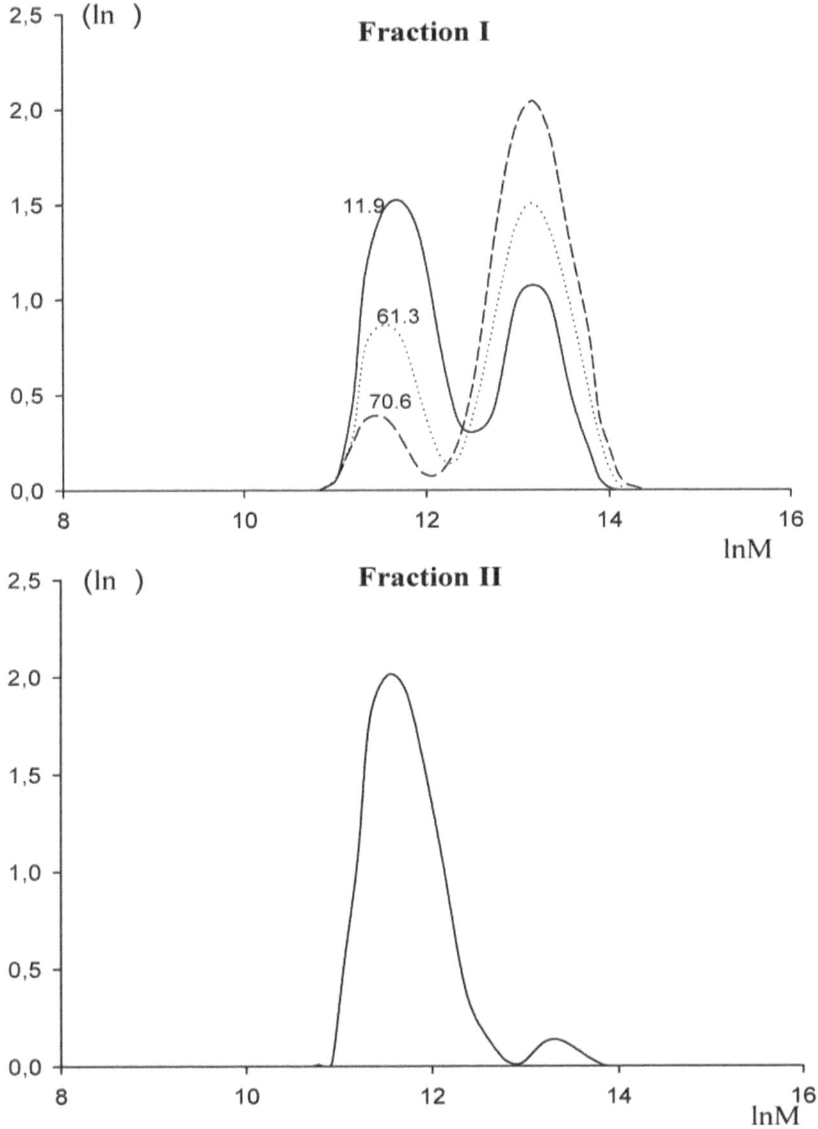

FIGURE 3.4 Active site distributions over kinetic heterogeneity during isoprene polymerization on fractions C-2, Method 1.

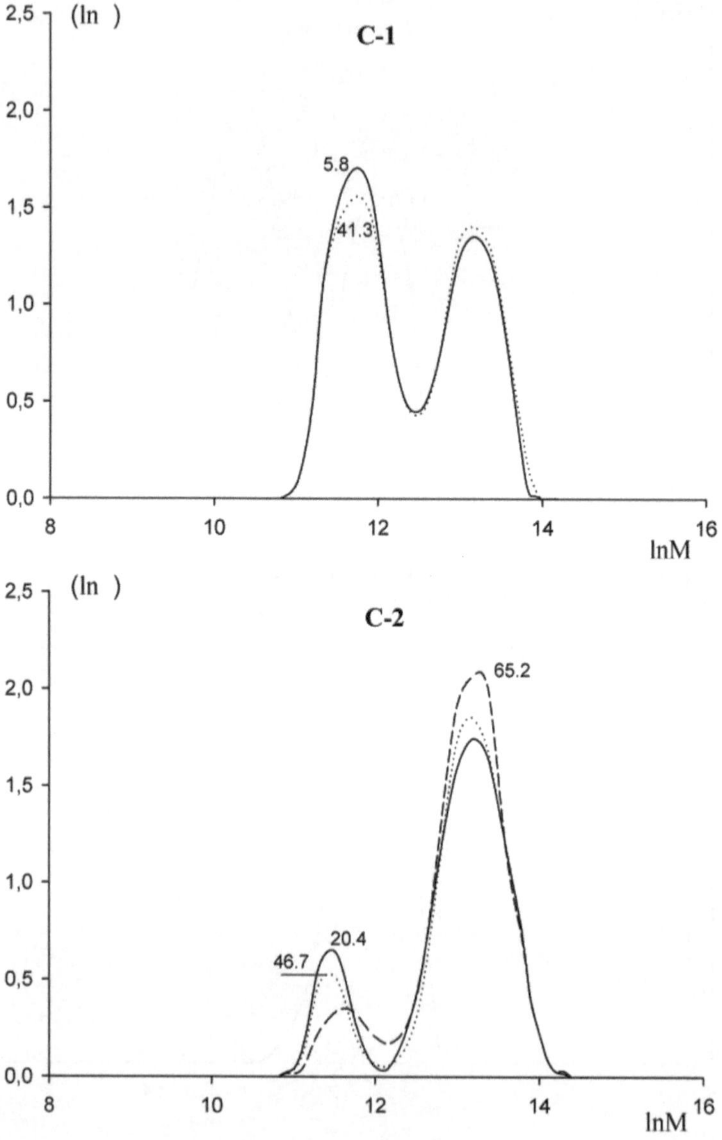

FIGURE 3.5 Active site distributions over kinetic heterogeneity during isoprene polymerization on particles of Fraction II of C-1 and C-2, Method 2.

In the presence of piperylene, the main distinction of function ψ(lnβ) relative to the distributions considered above is a significantly decreased area of the peak due to the active site of type B (Fig. 3.6). The decrease of the catalyst exposition temperature to −10°C (C-4, method 1) allows the complete "elimination" of active site of type B (Fig. 3.6). With allowance for the low content of fraction III, it may be concluded that, under these conditions, a single site catalyst is formed. In the case of the hydrodynamic action on C-4, the particles of fraction II contain active centers of type C with some shift of unimodal curve ψ(lnβ) to smaller molecular masses (Fig. 3.6).

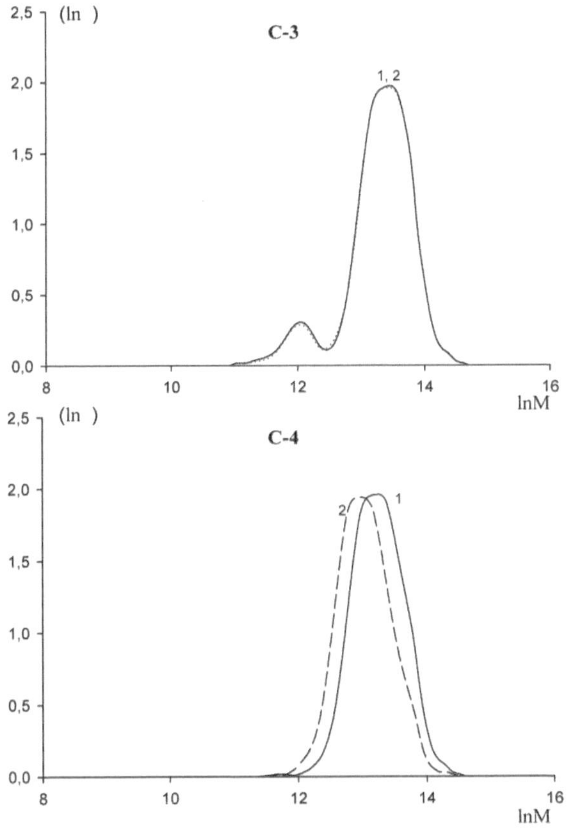

FIGURE 3.6 Active site distributions over kinetic heterogeneity during isoprene polymerization on particles of fraction II of C-3 and C-4. Methods 1–1, Method 2–2.

3.4 DISCUSSION

The particles of the titanium catalyst 0.03–0.14 μm in diameter, regardless of the conditions of catalytic system preparation, feature low activity in polyisoprene synthesis, and the resulting polymer has a low molecular mass and a narrow MWD. The molecular-mass characteristics of polyisoprene and the activity of the catalyst comprising particles 0.15–4.50 μm in diameter depend to a great extent on its formation conditions.

As shown in Refs. [10, 11], the region of coherent scattering for particles based on $TiCl_3$ spans 0.003–0.1μm, a range that corresponds to the linear size of the minimum crystallites. Coarser catalyst particles are aggregates of these minimum crystallites. This circumstance makes it possible to suggest that the fraction of catalyst particles 0.03–0.14μm in diameter that was isolated in this chapter is a mixture of primary crystallites of β-$TiCl_3$ that cannot be separated via sedimentation. The fractions of catalyst particles with larger diameters are formed by stable aggregates of 2–1100 primary crystallites. There is sense in the suggestion that the elementary crystallites are combined into larger structures via additional Al–Cl bonds between titanium atoms on the surface of a minimum of two elementary crystallites, that is, $(Ti)_1$–Cl–Al–Cl–$(Ti)_2$. Similar structures can be formed with the participation of AlR_2Cl and $AlRCl_2$, which are present in the liquid phase of the catalyst. Tri alkyl aluminum AlR_3 is incapable of this type of bonding. Thus, the structure of the most alkylated Ti atom (in the limit, a monometallic center of polymerization), which has the minimum reactivity, should be assigned to the active centers localized on particles 0.03–0.14 μm in diameter [9]. On particles 0.15–4.50μm in diameter in clusters of primary crystallites, high activity bimetallic centers with the minimum number of Ti–C bonds at a Ti atom are present. Thus, the experimental results obtained in this chapter show that the nature of the polymerization center resulting from successive parallel reactions between the pristine components of the catalytic system determines the size of the titanium catalyst particles and, consequently, their activity in isoprene polymerization.

3.5 CONCLUSION

We first examined the isoprene polymerization on the fractions of the titanium catalyst particles, which were isolated by sedimentation of the total mixture. The results obtained allow to consider large particle as clusters, which are composed of smaller particles. In the formation of these clusters are modified ligands available titanium atoms. In the process of polymerization or catalyst preparation the most severe effects are large particles (clusters). This result in the developing process later on substantially smaller as compared to initial size particles. These particles are fragments of clusters, which are located over the active centers of polymerization. Note that the stereo specificity is not dependent on the size of the catalyst particles.

Hypothesis about clusters agrees well with the main conclusions of this chapter:

1. Isolated the fraction of particles of titanium catalyst $TiCl_4$–Al (iso-$C_4H_9)_3$: I—0.7–4.5 μm, II—0.15–0.68 μm, III—0.03–0.13 μm. With decreasing particle size decreases the rate of polymerization the molecular weight and width of the molecular weight distribution. Hydrodynamic impact leads to fragmentation of large particles of diameter greater than 0.5 μm.
2. Isoprene polymerization under action of titanium catalyst is occurs on three types active sites: type A—lnM = 10.7; type B—lnM = 11.6; type C—lnM = 13.4. Fractions I and II particles contain the active site of type B and type C. The fraction III titanium catalyst is represented by only one type of active sites producing low molecular weight polymer (lnM = 10.7).
3. The use of hydrodynamic action turbulent flow, doping DPO and piperylene, lowering temperature of preparation of the catalyst allows to form single site catalyst with high reactivity type C (lnM = 13.4), which are located on the particles of a diameter of 0.15–0.18μm.

ACKNOWLEDGMENTS

This chapter was financially supported by the Council of the President of the Russian Federation for Young Scientists and Leading Scientific Schools Supporting Grants (Project No.: MD-4973.2014.8).

KEYWORDS

- **Active sites**
- **Isoprene polymerization**
- **Particles size effect**
- **Single site catalysts**
- **Ziegler-Natta catalyst**

REFERENCES

1. Kissin, Yu. V. (2012). *Journal of Catalysis, 292*, 188–200.
2. Hlatky, G. G. (2000). *Chemical Reviews, 100,* 1347–1376
3. Kamrul Hasan, A. T. M., Fang, Y., Liu, B., & Terano, M. (2010). *Polymer, 51*, 3627–3635.
4. Schmeal, W. R., & Street, J. R. (1972). *Journal of Polymer Science: Polymer Physics Edition, 10*, 2173–2183
5. Ruff, M., & Paulik, C. (2013). *Macromolecular Reaction Engineering, 7*, 71–83
6. Taniike, T., Thang, V. Q., Binh, N. T., Hiraoka, T., Uozumi, Y., & Terano, M. (2011). *Macromolecular Chemistry and Physics, 212*, 723–729.
7. Morozov, Yu. V., Nasyrov, I. Sh., Zakharov, V. P., Mingaleev, V. Z., & Monakov, Yu. B. (2011). *Russian Journal of Applied Chemistry, 84*, 1434–1437.
8. Zakharov, V. P., Berlin, A. A., Monakov, Yu. B., & Deberdeev, R. Ya. (2008). *Physicochemical Fundamentals of Rapid Liquid Phase Processes,* Moscow: Nauka, 348 p.
9. Monakov, Y. B., Sigaeva, N. N., & Urazbaev, V. N. (2005). *Active Sites of Polymerization. Multiplicity: Stereospecific and Kinetic Heterogeneity,* Leiden: Brill Academic, 397 p.
10. Grechanovskii, V. A., Andrianov, L. G., Agibalova, L. V., Estrin, A. S., & Poddubnyi, I. Ya. (1980). Vysokomol. *SoedinSer. A 22*, 2112–2120.
11. Guidetti, G., Zannetti, R., Ajò, D., Marigo, A., & Vidali, M. (1980). *European Polymer Journal, 16,* 1007–1015.

CHAPTER 4

TRENDS IN POLYBLEND COMPOUNDS PART 1

A. L. IORDANSKII, S. V. FOMIN, A. BURKOVA, YU. N. PANKOVA, and G. E. ZAIKOV

CONTENTS

ABSTRACT

Design of bioerodible composites combining synthetic and natural polymers is an important stage for development in constructional and packaging materials which are friendly environmental ones during exploitation and thrown on a garbage dump. For this object the composites on the base of poly(3-hydroxybutyrate) and polyisobutylene were prepared in plunger extruder. Morphology of the blends was studied by AFM (NTegra Prima) with cantilever NSG-01 with frequency 0.5–1.0 Hz and DSC (DSC-60–9Shimadzy, Japan). Rheology characteristics were obtained by plastometer Stress Tech (Rheological Instruments, Sweden). Degradation was performed in soil at 22–25 °C with following weighing to determine weight loss.

Two-phase matrices of the composites show displacement of glass transition temperatures of individual phases. Comparison of experimental and theoretical values of viscosity confirmed the formation of a continuous matrix with less viscous PHB. This greatly simplifies the processing of melts of compositions PHB-PIB. Microbial degradation of PHB phase occurs almost entirely within 100–125 days. The remaining friable matrixes PIB are much more susceptible to degradation by other destructive factors (oxygen of the air, temperature, mechanical stress). Process of biodegradation of PHB phase influences on the structure of PIB. This is reflected in a sharp decrease of average molecular weight of PIB.

4.1 INTRODUCTION

One of the most pressing problems facing humanity is increasing pollution. In this regard, question of recycling of synthetic polymeric materials arises very sharply. Plastics production rates are growing exponentially and production volumes hundreds of millions of tons annually. One of the most dynamic areas of the use of plastics is packaging industry (40 to 50% of the total production of plastics). Thus, billions of tons of municipal solid waste products constitute more than half of the short-term or one-time application on the basis of large common polyolefin. A possible solution of this problem is to create a biodegradable polymer composition. The fastest growing trend in this area is the use of polyhydroxyalkanoates (PHAs). Some physical and chemical characteristics of PHA are similar to these of synthetic polymers (polypropylene, polyethylene). However, in addition to their thermo plasticity, representatives of PHAs have optical activity, increase induction period of oxidation, exhibit the piezoelectric effect and, what is most important, they are characterized as being biodegradable and biocompatible. At the same time, the PHAs have disadvantages (high cost, brittleness), which can be partially or completely compensated by using composite materials based on blends with other polymers, with dispersed fillers or plasticizers. Taking into account all the above, we have suggested to create a mixed polymer composite based on poly 3-hydroxybutyrate (PHB) and polyisobutylene (PIB).

4.2 OBJECTS AND METHODS

The objects of the chapter were high molecular weight PIB of mark "P-200" and PHB Lot 16F. The poly 3-hydroxybutyrate was obtained by microbiological synthesis in company "BIOMER®" (Germany). PHB is a white fine powder; the density is 1.25 g/cm^3; the molecular weight is 325 kDa. Polyisobutylene of high molecular weight "P-200" is a white elastic material, transparent in thin films, odorless, with density of 0.93 g/cm^3 and with molecular weight of about 175–225 kDa.

These materials have been chosen due to their economic expediency and the valuable combination of physical and chemical, physical and mechanical and other properties of individual polymers – PHB is a brittle thermoplastic; PIB – an elastomer. Preparation of composite materials based on combinations of plastics and elastomers is well known: plastics are used as polymeric fillers in elastomers, improving their technological and working characteristics; elastomers effectively improve strike viscosity and reduce brittleness in compositions based on plastics.

The polymer compositions were prepared in the following proportions: PHB: PIB = 10:90, 20:80, 30:70, 40:60, 50:50 and 60:40 (by weight ratio here and hereinafter). Then, the individual polymers were investigated. The first stage of blending was carried out on laboratory mixing rolls with the roll length of 320 mm and diameter of 160 mm. Temperature of the back roll was 60 °C, temperature of the front roll was 50 °C. In these conditions the compositions were obtained, in which the micro powder of PHB was distributed in the continuous matrix of PIB. Deeper joining, mutual segmental solubility of polymers require an additional high temperature treatment of compositions, so the second stage of the composition processing was carried out in plunger extruder at the temperature of 185 °C.

Average molecular weight of the elastomer was characterized by viscometry method. Average molecular weight of PIB was calculated according to the equation Mark-Houwink:

$$[\eta] = K \times Mm^{\alpha}, \quad (1)$$

where [η] – the intrinsic viscosity, mL/g; Mm – average molecular weight of the polymer; K, α – constants for a given system "polymer-solvent."

Heptane was used as solvent, the constants $K=1.58\ 10^{-4}$, $\alpha=0.69$ [8]. For each sample, a series of solutions were prepared with different concentrations (0.2, 0.4, 0.6, 0.8 and 1.0 g/100 mL), and then determines the relative, specific and intrinsic viscosity. The calculated molecular weights presented in Table 4.1.

Investigation of the structure of the compositions by atomic force microscopy was performed on the tunnel atomic force microscope brands Ntegra Prima (company "NT-MDT") (cantilever NSG–01 with a frequency of 0.5–1 Hz). Investigations were carried out by semi contact mode. Microtome cuts were made on the brand Microm HM-525 (company "Thermo scientific," Germany).

Investigation of the structure of the compositions was carried out by differential scanning calorimetry on differential scanning calorimeter DSC–60 (company "Shimadzu," Japan). Samples of PHB-PIB blends weighing several mg were placed in open aluminum crucibles with a diameter of 5.8 mm and a height of 1.5 mm and weighing 13 mg upper temperature limit of 600 °C. Temperature range from minus 100 to plus 250 °C, heating rate of 10 deg/min. Liquid nitrogen is used to generate low temperature. Instrument calibration was performed according to indium, tin and lead.

The rheological curves of pure polymer melts and their mixtures were obtained with the multifunction rheometer "Stress Tech" (company "REOLOGICA Instruments AB," Sweden). The measuring cell consisted of two parallel planes (the lower plane was fixed, the upper plane being a rotating rotor); the shear rate was 0.1 s^{-1}, the temperature range was 443–513K (170–240 °C). The lower limit of the temperature range was chosen by the melting temperature of PHB (174.4 °C the data obtained for DSC), the upper limit is the beginning of irreversible degradation in polymers.

Polymer samples were subjected to soil degradation in a laboratory at a temperature 22–25°C. Samples in the form of a film thickness of 50 microns was placed in the soil to a depth of 1.5–2 cm. Biodegradation rate was assessed by evaluating the mass loss of the samples. Mass losses were fixed by weighing the samples on an analytical balance.

4.3 RESULTS AND DISCUSSION

Properties of mixed polymer compositions are determined by many factors, among which in the first place should be allocated phase structure (ratio and the size of the phase domains). Therefore, at the first stage of the research attention has been paid to study the structure of formed compositions. In the investigation of samples with a low content of PHB (10–30% by weight) has been found that it forms a discontinuous phase, that is, distributed in a continuous matrix PIB as separate inclusions of the order of 1–2 microns. The results of atomic force microscopy for the composition ratio of PHB-PIB 20:80 shown in Fig. 4.1.

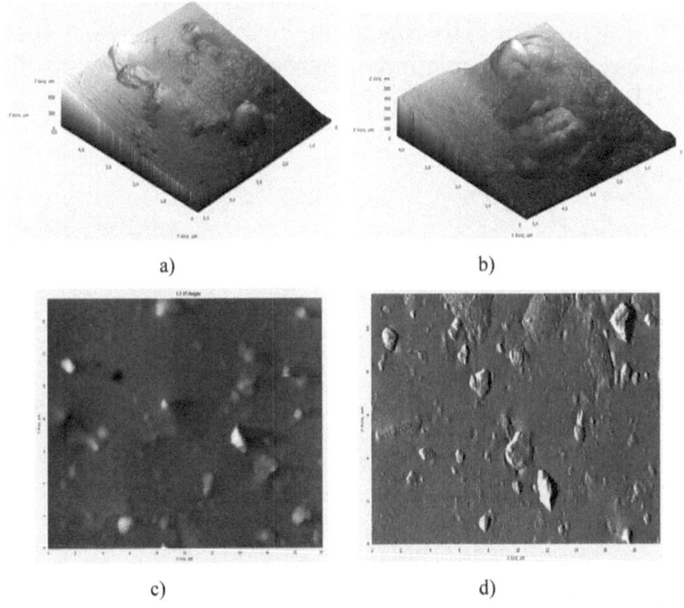

FIGURE 4.1 Photomicrographs of the relief of films with PHB-PIB ratio of 20:80 (scan size 5 × 5 um) (a, b) and AFM phase contrast image of the film with a ratio of 20:80 PHB-PIB (scan size 20 × 20 um) (c, d).

Atomic force microscopy is considered one of the most perspective methods for studying polymer blends is because this method allows to clearly define the phase boundary and its scale. Polymer identification was performed by controlling the interaction of the probe with the sample surface at different points. When approaching the surface of the cantilever is deflected downward (to the sample) due to attractive forces until the probe comes into contact with the sample. When the probe is withdrawn from the studied surface, a hysteresis is observed, associated with the adhesive forces. Adhesion forces between the probe and the sample are forcing them to remain in contact, which causes the cantilever to bend. Phase of the polymers were very clearly identified by mapping curves "approach-removal" of the probe.

Probable range of the phase inversion is an important characteristic for mixed biodegradable composites. This allows from a practical point of view to establish the minimum concentration of PHB (at which a continuous phase is formed) for intensive biodegradation as microorganisms must be able to penetrate deep into the mixed composite. When compared with the atomic force microscopy of samples with different proportions of the components was found that in investigated materials continuous matrix formation occurs when PHB content in the mixture of about 40–50% by weight. The results of atomic force microscopy for the composition ratio of PHB-PIB 50:50 shown in Fig. 4.2.

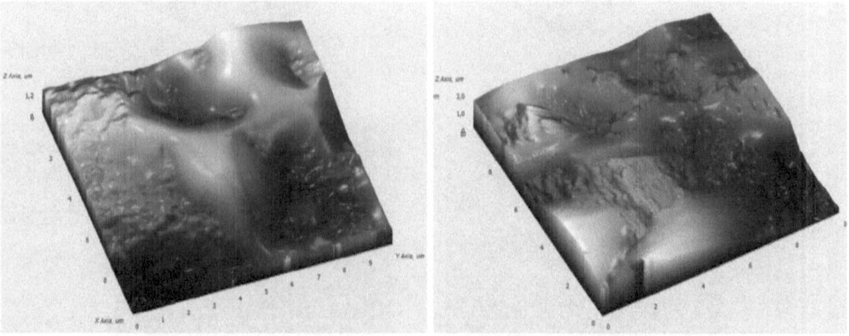

a) b)

FIGURE 4.2 *(Continued)*

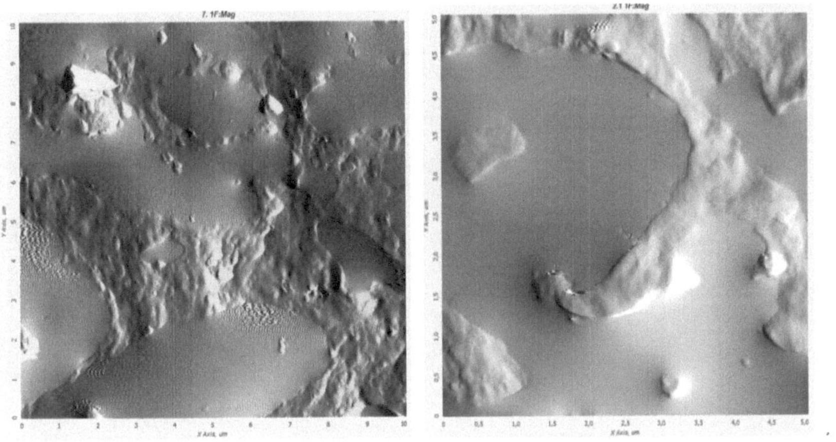

c) d)

FIGURE 4.2 Photomicrographs of the relief of films with PHB-PIB ratio of 50:50 (scan size 10 × 10 um) (a, b) and AFM phase contrast image of the film with a ratio of 50:50 PHB-PIB (scan size 10 × 10 um) (c) and 5 × 5 um) (d).

Thus, by atomic force microscopy were determined scale structures and distributions of polymers according to the ratio of the starting components in the mixture, and the approximate range of probable phase inversion (about 40–50% by weight of PHB).

Determination of the glass transition temperature is an informative method of research the phase structure of polymer blends. In the case of mixed compositions glass transition may occur in each phase separately, if polymers do not interact with each other. In another case glass transition in mixtures is fully cooperative process, involving macromolecules mixed polymer segments. This composition would have single glass transition temperature, which varies monotonically depending on the mixture composition [1]. The glass transition temperatures of polymer components of the mixtures were determined by differential scanning calorimetry. The results of DSC shown in Fig. 4.3.

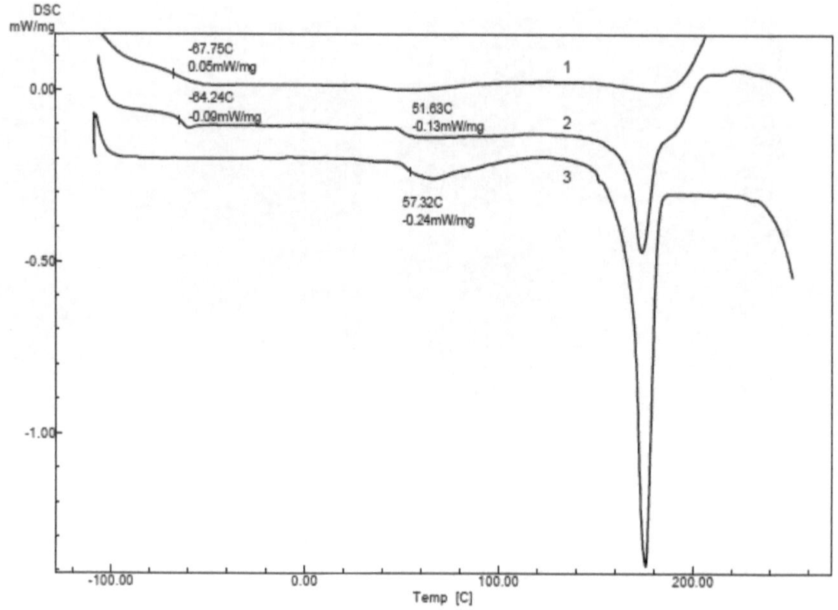

FIGURE 4.3 DSC curves of the samples (1 – pure PIB, 2–20% PHB, 3–pure PHB).

In all cases, the specific heat jump was recorded, which corresponds to the glass transition of phases polymers. It was noted displacement values of glass transition temperature of phase PIB to higher temperatures with increasing content of PHB (from 68°C for pure PIB to 64°C for the composition with 60% by weight of the PHB). For PHB phase displacement of values of glass transition temperature is shown in Fig. 4.4.

As can be seen from the data presented, the displacement of values of glass transition temperature of PHB occurs about 6–7°C. Most often in the literature as reasons for this phenomenon is called a limited solubility of the mixture components in each other (from a fraction of a few percent) [1–3]. However, for the polymers probability such variant is extremely small because of the significant thermodynamic incompatibility. More probable reasons of displacement of the glass transition temperature may be changes of the supramolecular structure of polymers when mixing; as well as differences in thermal expansion coefficients of polymers in the region above and below the glass transition temperature.

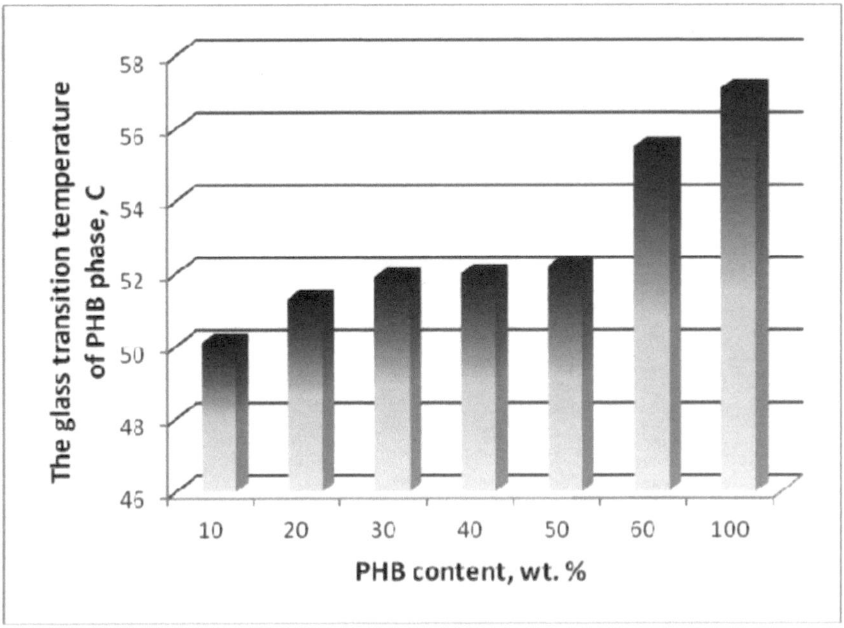

FIGURE 4.4 Changes in the glass transition temperature of phase PHB.

The glass transition temperature change deserves attention because is a sharp jump in the of PHB content of 50–60% by weight. This phenomenon may also be due to the probable phase inversion (previously established by microscopy methods) at said ratio of the components. Continuous structure more rigid of PHB has a higher glass transition temperature than the individual inclusions of thermoplastic material, isolated from each other by a continuous matrix of PIB. Thus, as a result of studies on the structure of compositions of PHB-PIB was confirmed formation of heterogeneous two-phase systems. For biodegradable polymer compositions it may be advantage, because this system is more susceptible to external influences destructive.

However, for practical use of PHB-PIB compositions is necessary to evaluate the possibility of processing these materials into finished products. As most polymer blends are processed by melting them, investigations of rheological properties of these compositions are of great scientific and practical interest. Information about the structure of composition that can be obtained on the basis of rheological investigation is the level of intermolecular interactions, the degree of macromolecules ordering, the phase structure of polymer blends. Dependence "melt viscosity temperature" were investigated for all compositions. The viscosity values at various temperatures for blends of polymers with different ratios are shown in Fig. 4.5.

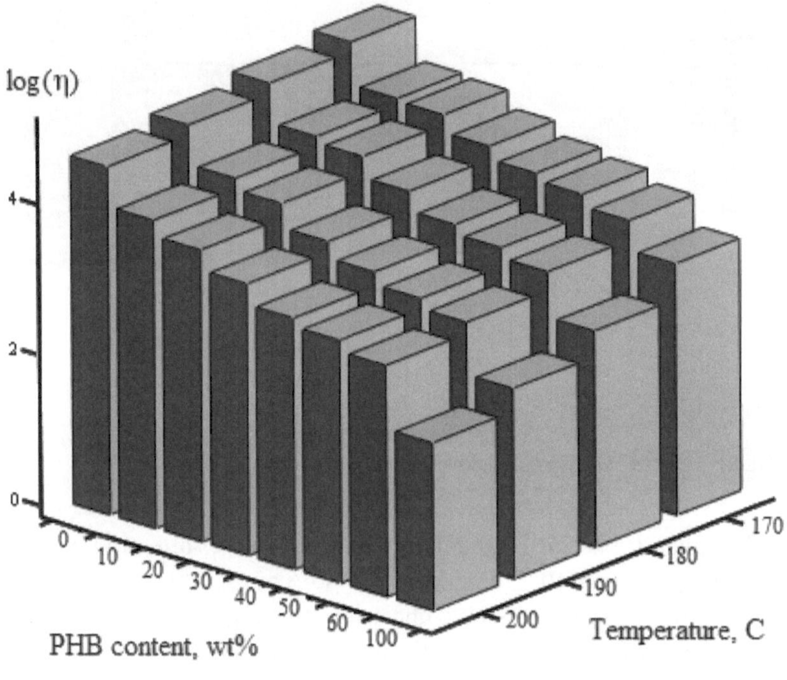

FIGURE 4.5 The viscosity values at various temperatures for blends of polymers with different ratios.

The viscosity values of the composition decreases considerably with increasing temperature and increasing the content of PHB that gives the chance regulation of technological properties of compositions in a wide range. Based on the significant difference of the solubility parameters calculated previously [6], it seems reasonable assumption on the formation of a biphasic system melts of mixtures of these polymers. The assumption of formation of two-phase mixtures of melts of these polymers seems justified based on significant difference in the solubility parameters, which calculated previously [6]. Kerner-Takayanagi equation applicable to describe the rheological properties of two-phase mixtures of polymers [1]:

$$\eta = \eta_m \times \frac{(7-5v_m)\times \eta_m + (8-10v_m)\times \eta_f - (7-5v_m)\times(\eta_m - \eta_f)\times f_f}{(7-5v_m)\times \eta_m + (8-10v_m)\times \eta_f - (8-10v_m)\times(\eta_m - \eta_f)\times f_f}, \quad (2)$$

where η – viscosity of the mixture, Pa × s; η_m – viscosity of the matrix, Pa × s; η_f – viscosity of the dispersed phase, Pa × s; φ_f – volume fraction of the dispersed phase; ν_m – Poisson's ratio of the matrix (assumed equal to 0.5).

This expression describes the dependence of the properties of the mixture composition, excluding inevitable phase inversion in the mixture. According to Eq. (2) the viscosity of the blend composition is graphically expressed by the two curves corresponding to the two limiting cases when one or the other phase is continuous throughout the range of compositions of the polymer mixture. Theoretical values of the viscosity of the compositions of various compositions were calculated based on the viscosity base polymers. Calculations on model of Kerner-Takayanagi at different temperatures are shown in Fig. 4.6.

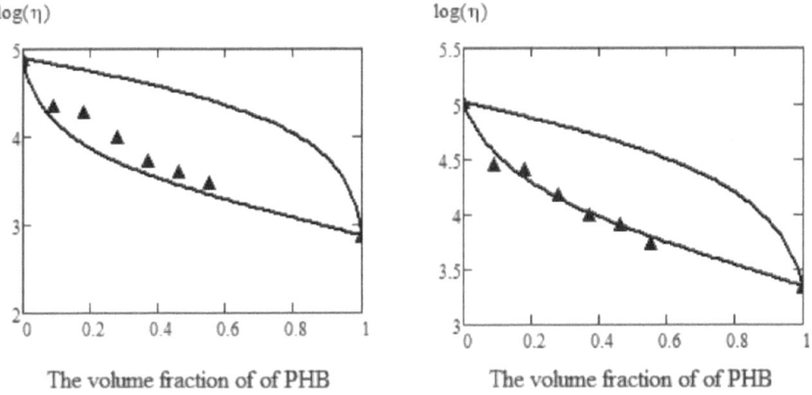

FIGURE 4.6 Viscosity of the composition at a temperature of 180°C (a) and 170°C (b) (theoretical calculations on model of Kerner-Takayanagi; ▲ – experimental data).

By analyzing data of Figure 6, it was found that PHB forms a continuous matrix in the molten polymer at any ratio. The experimental values of the viscosity are close to the bottom theoretical curve, which corresponds to the calculation for the case of the formation of the matrix of PHB. Thus, during the melting of PHB observed phase inversion phenomenon in accordance with the laws of more fluid melt [4, 6] PHB forms a continuous phase in the entire range of concentrations. Basic rheological parameters of polymer blends determined by the properties of the polymer matrix, so the properties of melts are generally defined by PHB. Lower viscosity of the melt blend composition (due to the formation of a continuous matrix PHB) in this case will significantly simplify the processing of the investigated materials. Also, the data in Fig. 4.6 are consistent with the position that the greater the difference in viscosity of mixed polymers, the earlier the formation of a continuous matrix less viscous component.

Previous studies [7] revealed that the main physical and mechanical characteristics (tensile strength, tensile modulus, elongation at break) PHB-PIB composition containing 30–50% by weight of PHB, not inferior to traditional polymers used in the production of packaging (polyethylenes of different brands). But the undeniable advantage of the compositions PHB-PIB is the possibility of biodegradation. It has been found that mixtures containing 40–60% PHB stepwise lose mass when exposed to soil. In the first stage (100 days) intensive degradation of phase PHB occurs under the action of microorganisms. The result of the first stage of this scenario is PIB samples with very high surface area. In the second step (after 100 days), mass loss is stabilized and is much slower.

TABLE 4.1 The Average Molecular Weight of PIB Samples

PIB samples	The intrinsic viscosity	The average molecular weight
The original sample	0.68	185–000
The sample after 200 days of exposure in the soil (0% PHB)	0.65	175–000
The sample after 200 days of exposure in the soil (60% PHB)	0.49	115–000

As can be seen from Table 4.1, the average molecular weight PIB sample (previously containing PHB) much less compared to the original molecular weight PIB and PIB after exposure to soil. Significant impact of PHB phase on the ability of PIB to degradation in soil was found by research of the molecular weight of the elastomer. Destruction of the original structure PIB can occur due to physical factors (large specific surface films), chemical factors (interaction of the products of biodegradation of PHB with PIB phase, and initiation of the degradation of the elastomer) and biological factors (accumulation of microorganisms biomass on the surface of mixed films and therefore some microorganisms can use the matrix PIB as a carbonaceous substrate for growth.

4.4 CONCLUSIONS

There are some conclusions drawn in the course of studying structure and properties of polymer system "PHB-PIB":

1. Structure of mixed compositions of PHB-PIB investigated by atomic force microscopy; confirmed the formation of a heterogeneous two-phase structure. PHB forms a continuous elongated structure in the matrix of PIB when content of PHB in the mixture is greater than 30%.
2. Displacement of glass transition temperatures of individual phases in a mixture of polymers determined by differential scanning calorimetry. This may be indicative of their limited interaction, despite the significant difference of the solubility parameters.
3. Comparison of experimental and theoretical values of viscosity confirmed the formation of a continuous matrix less viscous PHB. This greatly simplifies the processing of melts of compositions PHB-PIB.
6. Microbial degradation of PHB phase occurs almost entirely within 100–125 days. The remaining friable matrixes PIB are much more susceptible to degradation by other destructive factors (oxygen of the air, temperature, mechanical stress).
7. Process of biodegradation of PHB phase influences on the structure of PIB. This is reflected in a sharp decrease of average molecular weight of PIB.

ACKNOWLEDGMENTS

The work was supported by the Russian Foundation for Basic Research (grant no. 13–03–00405-a) and the Russian Academy of Sciences under the program "Construction of New Generation Macromolecular Structures" (03/OC-14).

KEYWORDS

- **AFM**
- **Blends**
- **Degradation**
- **DSC**
- **Morphology**
- **Poly(3-hydroxybutyrate)**
- **Polyisobutylene**
- **Rheology**
- **Soil**
- **Weight loss**

REFERENCES

1. Kuleznev, V. N. (1980). *Mixtures of the polymers*. Moscow.
2. Paul, D., & Bucknell, K. (2009). *Polyblend Compounds*. Functional Properties. Saint Petersburg.
3. Paul, D., & Bucknell, K. Polyblend. (2009). *Compounds. Systematics*; Saint Petersburg.
4. Malkin, A. (2007). *Rheology: Concepts, Methods, and Applications*. St. Petersburg.
5. Iordanskii, A. L., Fomin, S. V., & Burkov, A. A. (2012). Structure and Melt Rheology Mixed Compositions of Polyisobutylene and Poly 3-Hydroxybutyrate. *Plastic Materials, 7(7),* 13–16.
6. Schramm, G. (2003). *Basics Practical Rheology and Rheometry*. Moscow.
7. Iordanskii, A. L., Fomin, S. V., & Burkov, A. A. (2013). Investigation of the Structure and Properties of Biodegradable Polymer Composites Based on Poly 3-Hydroxybutyrate and Polyisobutylene. *Bulletin of the Kazan University of Technology, 16(9)*, 115–119.
8. Sangalov, Y. A. (2001). *Polymers and Copolymers of Isobutylene*. Fundamental and Applied Aspects of the Problem, Ufa.

CHAPTER 5

TRENDS IN POLYBLEND COMPOUNDS PART 2

A. P. BONARTSEV, A. P. BOSKHOMODGIEV, A. L. IORDANSKII,
G. A. BONARTSEVA, A. V. REBROV, T. K. MAKHINA,
V. L. MYSHKINA, S. A. YAKOVLEV, E. A. FILATOVA,
E. A. IVANOV, D. V. BAGROV, G. E. ZAIKOV, and M. I. ARTSIS

CONTENTS

5.1 INTRODUCTION

This chapter is designed to be an informative source for biodegradable poly(3-hydroxybutyrate) and its derivatives' research. We focuses on hydrolytic degradation kinetics at 37 °C and 70 °C in phosphate buffer to compare PLA and PHB kinetic profiles. Besides, we reveal the kinetic behavior for copolymer PHBV (20% of 3-hydroxyvalerate) and the blend PHB-PLA. The intensity of biopolymer hydrolysis characterized by total weight lost and the viscosity-averaged molecular weight (MW) decrement. The degradation is enhanced in the series PHBV < PHB < PHB-PLA blend < PLA. Characterization of PHB and PHBV includes MW and crystallinity evolution (X-ray diffraction) as well as AFM analysis of PHB film surfaces before and after aggressive medium exposition. The important impact of MW on the biopolymer hydrolysis is shown.

The bacterial polyhydroxyalkanoates (PHA)s and their principal representative poly(3-R-hydroxybutyrate) (PHB) create a competitive option to conventional synthetic polymers such as polypropylene, polyethylene, polyesters, etc. These polymers are nontoxic and renewable. Their biotechnology output does not depend on hydrocarbon production as well as their biodegradation intermediates and resulting products (water and carbon dioxide) do not provoke the adverse actions in environmental media or living systems [1–3]. Being friendly environmental [4], the PHB and its derivatives are used as the alternative packaging materials, which are biodegradable in the soil or different humid media [5, 6].

The copolymerization of 3-hydroxybutyrate entities with 3-hydroxyoctanoate (HO), 3-hydroxyheptanoate (HH) or 3-hydroxyvalerate (HV) monomers modifies the physical and mechanical characteristics of the parent PHB, such as ductility and toughness to depress its processing temperature and embrittlement. Besides, copolymers PHB-HV [7], PHB-HH [8] or PHB-HO [9], etc, have improved thermo physical and/or mechanical properties and hence they expand the spectrum of constructional and medical materials/items. For predicting the behavior of PHB and its copolymers in a aqueous media, for example, in vitro, in a living body or in a wet soil, it is essential to study kinetics and mechanism of hydrolytic destruction.

Despite the history of such-like investigations reckons about 25 years, the problem of (bio)degradation in semicrystalline biopolymers is too far from a final resolution. Moreover, in the literature the description of hydrolytic degradation kinetics during long-term period is comparatively uncommon [10–14]. Therefore, the main object of this chapter is the comparison of long-term degradation kinetics for the PLA, PHB and its derivatives, namely its copolymer with 3-oxyvalerate (PHBV) and the blend PHB/PLA. The contrast between degradation profiles for PHB and PLA makes possible to compare the degradation behavior for two most prevalent biodegradable polymers. Besides, a significant attention is devoted to the impact of molecular weight (MW) for above polymer systems upon hydrolytic degradation and morphology (crystallinity and surface roughness) at physiological (37 °C) and elevated (70 °C) temperatures.

5.2 EXPERIMENTAL PART

5.2.1 MATERIALS

In this work we have used poly L-lactide (PLA) with different molecular weights: 67, 152, and 400 kDa (Fluka Germany); chloroform (ZAO EKOS-1, RF), sodium valerate (Sigma-Aldrich, USA), and mono-substituted sodium phosphate (NaH_2PO_4, Chim Med, RF).

5.2.2 PHAS PRODUCTION

The samples of PHB and copolymer of hydroxybutyrate and hydroxyvalerate (PHBV) have been produced in A. N. Bach's Institute of Biochemistry. A highly efficient strain-producer (80 wt.% PHB in the dry weight of cells), *Azotobacter chroococcum* 7Б, has been isolated from rhizosphere of wheat (the sod-podzol soil). Details of PHB biosynthesis have been published in Ref. [15]. Under conditions of PHBV synthesis, the sucrose concentration was decreased till 30 g/L in medium and, after 10 h incubation, 20 mM sodium vale rate was added. Isolation and purification of the biopolymers were performed via centrifugation, washing and drying at 60 °C subsequently. Chloroform extractions of BPHB or BPHBV from the dry biomass and precipitation, filtration, washing again and drying have been described in Ref. [15]. The monomer-content (HB/HV ratio) in PHBV has been determined by nuclear magnetic resonance in accordance with procedure described in Ref. [16]. The percent concentration of HV moiety in the copolymer was calculated as the ratio between the integral intensity of methyl group of HV (0.89 ppm) and total integral intensity the same group and HB group (1.27 ppm). This value is 21 mol%.

5.2.3 MOLECULAR WEIGHT DETERMINATION

The viscosity-averaged molecular weight (MW) was determined by the viscosity (η) measurement in chloroform solution at 30 °C. The calculations of MW have been made in accordance with Mark-Houwink equation [17]:

$$[\eta] = 7.7 \cdot 10^{-5} \cdot M^{0.82}$$

5.2.4 FILM PREPARATIONS OF PHAS, PLA AND THEIR BLENDS

The films of parent polymers (PHB, PHBV and PLA) and their blends with the thickness about 40μm were cast on a fat-free glass surface. We obtained the set of films with different MW = 169±9 (defined as PHB 170), 349±12 (defined as PHB 350), 510±15 kDa (defined as PHB 500) and 950±25 kDa (defined as PHB 1000) as well as the copolymer PHBV with MW=1056±27 kDa (defined as PHBV). Additionally we prepared the set of films on the base of PLA with same thickness 40 μm and MW=67 (defined as PLA 70), MW=150 and 400kDa. Along with them we obtained the blend PHB/PLA with weight ratio 1:1 and MW = 950 kDa for PHB, and MW = 67 kDa for PLA (defined as PHB+PLA blend). Both components mixed and dissolved in common solvent, chloroform and then cast conventionally on the glass plate. All films were thoroughly vacuum-processed for removing of solvent at 40 °C.

5.2.5 HYDROLYTIC DEGRADATION IN VITRO EXPERIMENTS

Measurement of hydrolytic destruction of the PHB, PLA, PHBV films and the PHB-PLA composite was performed as follows. The films were incubated in 15 mL 25 mM phosphate buffer, pH 7.4, at 37 °C or 70°C in a ES 1/80 thermostat (SPU, Russia) for 91 days; pH was controlled using an Orion 420+ pH-meter (Thermo Electron Corporation, USA). For polymer weight measurements films were taken from the buffer solution every three day, dried, placed into a thermostat for 1 h at 40°C and then weighed with a balance. The film samples weighed 50–70 mg each. The loss of polymer weight due to degradation was determined gravimetrically using a AL-64 balance (Acculab, USA). Every three days the buffer was replaced by the fresh one.

5.2.6 WIDE ANGLE X-RAY DIFFRACTION

The PHB and PHBV chemical structure, the type of crystal lattice and crystallinity was analyzed by wide-angle X-ray scattering (WAXS) technique. X-ray scattering study was performed on device on the basis of 12 kW generator with rotating copper anode RU-200 Rotaflex (Rigaku, Japan) using CuK radiation (wavelength $\lambda = 0.1542$ nm) operated at 40 kV and 140 mA. To obtain pictures of wide angle X-ray diffraction of polymers two-dimensional position-sensitive X-ray detector GADDS (Bruker AXS, Germany) with flat graphite monochromator installed on the primary beam was used. Collimator diameter was 0.5 mm [18].

5.2.7 ATOMIC FORCE MICROSCOPY OF PHB FILMS

Microphotographs of the surface of PHB films were obtained be means of atomic force microscopy (AFM). The AFM imaging was performed with Solver PRO-M (Zelenograd, Russia). For AFM imaging a piece of the PHB film (~2×2 mm^2) was fixed on a sample holder by double-side adhesive tape. Silicon cantilevers NSG11 (NT-MDT, Russia) with typical spring constant of 5.1 N/m were used. The images were recorded in semi-contact mode, scanning frequency of 1–3 Hz, scanning areas from 3×3 to 20×20 μm^2, topography and phase signals were captured during each scan. The images were captured with 512×512 pixels. Image processing was carried out using Image Analysis (NT-MDT, Russia) and FemtoScan Online (Advanced technologies center) software.

5.3 RESULTS AND DISCUSSION

The in vitro degradation of PHB with different molecular weight (MW) and its derivatives (PHBV, blend PHB/PLA) prepared as films was observed by the changes of total weight loss, MW, and morphologies (AFM, XRD) during the period of 91 days.

5.3.1 THE HYDROLYSIS KINETICS OF PLA, PHB, AND ITS DERIVATIVES

The hydrolytic degradation of the biopolymer and the derivatives (the copolymer PHBV, and the blend PHB/PLA 1:1) has been monitored for three months under condition, which is realistically approximated to physiological conditions, namely, in vitro: phosphate buffer, pH=7.4, temperature 37 °C. The analysis of kinetic curves for all samples shows that the highest rate of weight loss is observed for PLA with the smallest MW ≈ 70 kDa and for PHB with relatively low MW ≈ 150 kDa (Fig. 5.1). On the base of the data in this figure it is possible to compare the weight-loss increment for the polymers with different initial MW. Here, we clearly see that the samples with the higher MWs (300–1000 kDa) are much stabler against hydrolytic degradation than the samples of the lowest MW. The total weight of PHB films with MW=150 kDa decreases faster compared to the weight reduction of the other PHB samples with higher MW's = 300 and 450 or 1000 kDa. Additionally, by the 91st day of buffer exposition the residual weight of the low-MW sample reaches 10.5% weight loss that it is essentially higher than the weight loss for the other PHB samples (Fig. 5.1).

After establishing the impact of MW upon the hydrolysis, we have compared the weight-loss kinetic curves for PLA and PHB films with the relatively comparative MW = 400 and 350 kDa, respectively and the same film thickness. For the PLA films one can see the weight depletion with the higher rate than the analogous samples of PHB. The results obtained here are in line with the preceding literature data [8, 12, 19–21].

Having compare destruction behavior of the homopolymer PHB and the copolymer PHBV, we can see that the introduction of hydrophobic entity (HV) into the PHB molecule via copolymerization reveals the hydrolytic stability of PHBV molecules. For PHBV an hydrolysis induction time is the longest among the other polymer systems and over a period of 70 days its weight loss is minimal (<1% wt.) and possibly related with desorption of low-molecular fraction of PHBV presented initially in the samples after biosynthesis and isolation. The kinetic curves in Fig. 5.1 show also that the conversion the parent polymers to their blend PHB-PLA decreases the hydrolysis rate compared to PHB (MW=1000 kDa) even if the second component is a readily hydrolysable polymer: PLA (MW=70 kDa).

For the sake of hydrolysis amplification and its exploration simultaneously, an polymer exposition in aqueous media has usually been carried out at elevated temperature [11, 19]. To find out a temperature impact on degradation and intensify this process, we have elevated the temperature in phosphate buffer to 70 °C. This value of temperature is often used as the standard in other publications see, for example Ref. [11]. As one should expect, under such condition the hydrolysis acceleration is fairly visible that is presented in Fig. 5.1b. By the 45th day of PLA incubation its films turned into fine-grinding dust with the weight-loss equaled 50% (MW=70 kDa) or 40% (MW=350 kDa). Simultaneously the PHB with the lowest MW=170 kDa has the weight loss = 38 wt.% and the film was markedly fragmented while the PHB samples with higher MWs 350, 500 and 1000 kDa have lost the less percent of the initial weight, namely 20, 15 and 10%, respectively. Additionally, for 83 days the weight drop in the PHB-PLA blend films is about 51 wt.% and, hence, hydrolytic stability of the blend polymer system is essentially declined (cf. Figs. 5.1a and 5.1b).

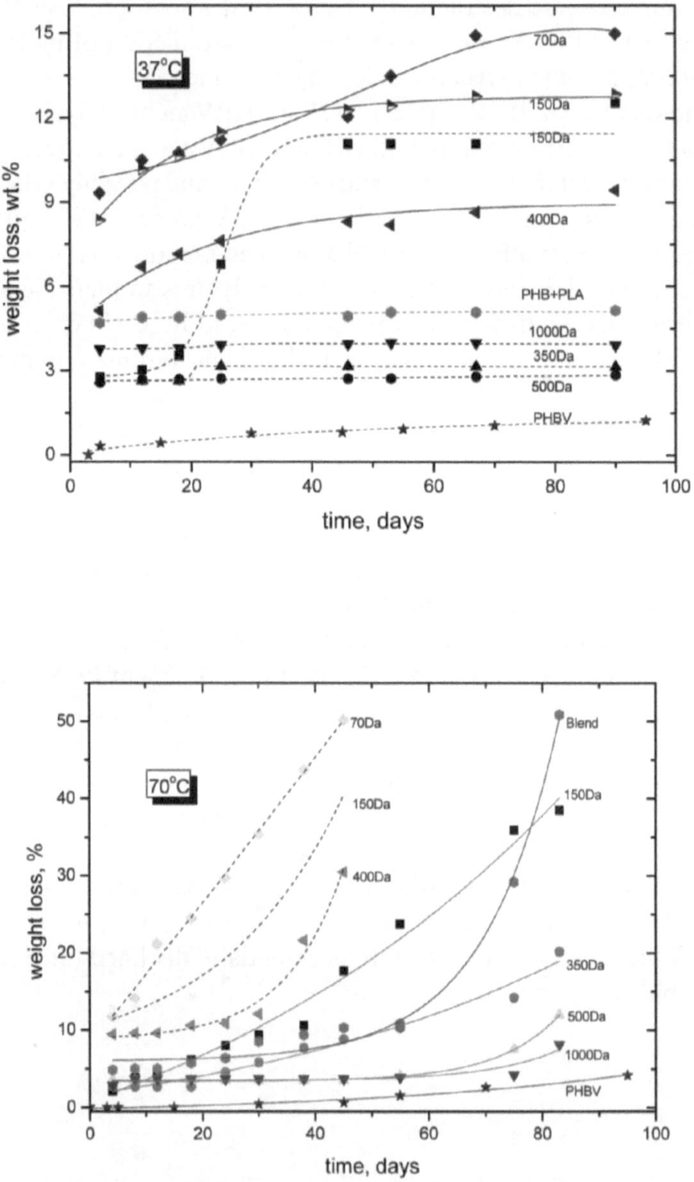

FIGURE 5.1 Weight loss in the phosphate buffer for PHB and its derivatives with different MW (shown on the curves in kDa). 37 °C, 70 °C: ♦, ▶, and ◀ are PLA films with MW=70, 150, and 400kDa, respectively; ■, ▲, ●, and ▼ are PHB samples with 170, 350, 500, and 1000 kDa, respectively; PHBV 1050 (★); and PHB-PLA blend (⬢).

At elevated temperature of polymer hydrolysis (70 °C) as well as at physiological temperature 37 °C we have demonstrated again that the PHBV films are the stablest because by 95th day they lost only 4 wt.%. The enhanced stability of PHBV relative to the PHB has been confirmed by other literature data [21]. Here it is worth to remark that during biosynthesis of the PHBV two opposite effects of water sorption acting reversely each other occur. On the one side, while the methyl groups are replaced by ethyl groups, the total hydrophobicity of the copolymer is enhanced, on the other side, this replacement leads to decrease of crystallinity in the copolymer [22]. The interplay between two processes determines a total water concentration in the copolymer and hence the rate of hydrolytic degradation. Generally, in the case of PHBV copolymer (HB/HV = 4:1 mol. ratio) the hydrophobization of its chain predominates the effect of crystallinity decrease from 75% for PHB to ~60% for PHBV.

5.3.2 CHANGE OF MOLECULAR WEIGHT FOR PHB AND PHBV

On exposure of PHB and PHBV films to buffer medium at physiological (37 °C) or elevated (70 °C) temperatures, we have measured both their total weight loss (Section 1) and the change of their MW simultaneously. In particular, we have shown the temperature impact on the MW decrease that will be much clear if we compare the MW decrements for the samples at 37° and 70 °C. At 70 °C the above biopolymers have a more intensive reduction of MW compared to the reduction at 37 °C (see Fig. 5.2). In particular, at elevated temperature the initial MW (= 350 kDa) has the decrement by 7 times more than the MW decrement at physiological condition. Generally, the final MW loss is nearly proportional to the initial MW of sample that is correct especially at 70 °C. As an example, after the 83-days incubation of PHB films, the initial MW= 170 kDa dropped as much as 18 wt.% and the initial MW= 350 kDa has the 9.1 wt.% decrease.

The diagrams in Fig. 5.2 shows that the sharp reduction of MW takes place for the first 45 days of incubation and after this time the MW change becomes slow. Combining the weight-loss (Section 1) and the MW depletion, it is possible to present the biopolymer hydrolysis as the two-stage process. On the initial stage, the random cleavage of macromolecules and the MW decrease without an significant weight-loss occur. Within this time the mean length of PHB intermediates is fairly large and the molar ratio of the terminal hydrophilic groups to the basic functional groups in a biodegradable fragment is too small to provide the solubility in aqueous media. This situation is true for the PHB samples with middle and high MW (350, 500, and 1000 kDa) when at 37 °C their total weight remains stable during all time of observation but the MW values are decreased till 76, 61, and 51 wt.%, respectively. On the second stage of degradation, when the MW of the intermediate molecules attains the some "critical" value and the products of hydrolysis become hydrophilic to provide dissolution and diffusion into water medium, the weight reduction is clearly observed at 70 °C. This stage is accompanied by the changes of physical-chemical, mechanical and structural characteristics and a geometry alteration. A similar 2-stage mechanism of PHB degradation has been described in the other publications [23, 24]. Furthermore, in the classical work of Reush [25] she showed that hydrophilization of PHB intermediates occurs at relatively low MW namely, at several decades of kDa. Our results provide evidences that the reduction of MW till "critical" values to be equal about 30 kDa leads to the expansion of the second stage, namely, to the intensive weight loss.

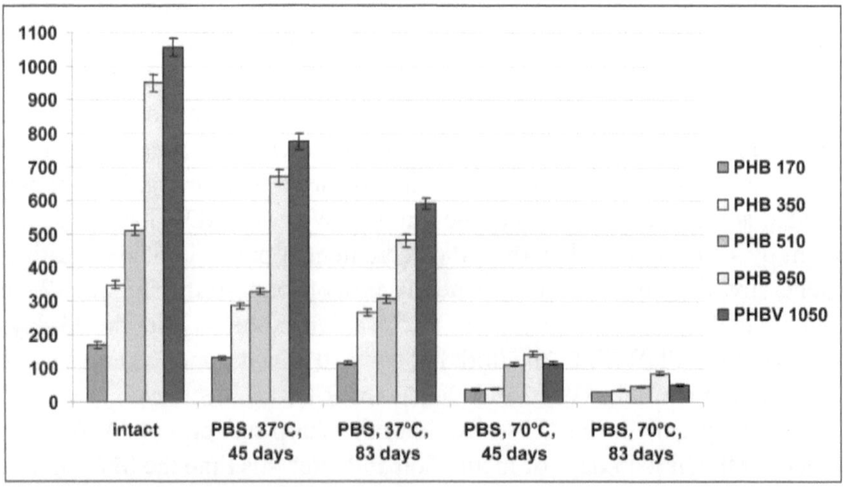

FIGURE 5.2 The molecular weight conversion of PHB and PHBV films during hydrolysis in phosphate buffer (PBS), pH = 7.4, 37 °C and 70 °C.

5.3.3 CRYSTALLINITY OF PHB AND PHBV

We have above revealed that during hydrolytic degradation, PHB and PHBV show the MW reduction (Section 2) and the total weight decrease (Section 1). Additionally, by the X-ray diffraction technique (XRD) we have measured the crystallinity degree of PHB and PHBV that varied depending on time in the interval of values 60–80% (see Fig. 5.3a). We have noted that on the initial stage of polymer exposition to the aqueous buffer solution (at 37 °C for 45 days) the crystallinity degree has slightly increased and then, under following exposition to the buffer, this characteristic is constant or even slightly decreased showing a weak maximum. When taken into account that at 37 °C the total weight for the PHB films with MWs equal 350, 500 and 1000 kDa and the PHBV film with MW equals 1050 are invariable, a possible reason of the small increase in crystallinity is recrystallization described earlier for PLA [26]. Recrystallization (or additional crystallization) happens in semicrystalline polymers where the crystallite portion can increase using polymer chains in adjoining amorphous phase [22].

At higher temperature of hydrolysis, 70 °C, the crystallinity increment is strongly marked and has a progressive trend. The plausible explanation of this effect includes the hydrolysis progress in amorphous area of biopolymers. It is well known that the matrices of PHB and PHBV are formed by alternative crystalline and noncrystalline regions, which determine both polymer morphologies and transport of aggressive medium. Additionally, we have revealed recently by H–D exchange FTIR technique that the functional groups in the PHB crystallites are practically not accessible to water attack. Therefore, the hydrolytic destruction and the weight decrease are predominantly developed in the amorphous part of polymer [22, 27]. Hence, the crystalline fraction becomes larger through polymer fragment desorption from amorphous phase. This effect takes place under the strong aggressive conditions (70 °C) and does not appear under the physiological conditions (37 °C) when the samples have invariable weight.

Owing to the longer lateral chains in PHBV, copolymerization modifies essentially the parent characteristics of PHB such as decreasing in crystallinity, the depression of melting and glass temperatures and, hence, enhancing ductility and improvement of processing characteristics [14, 28, 29]. Additionally, we have founded out that the initial crystallinity of PHB films is a monotonically increased function of initial MW (see Fig 5.3b). For samples with relatively low molecular weight it is difficult to compose the perfect crystalline entities because of a relatively high concentration of terminal groups performing as crystalline defects (Fig. 5.3).

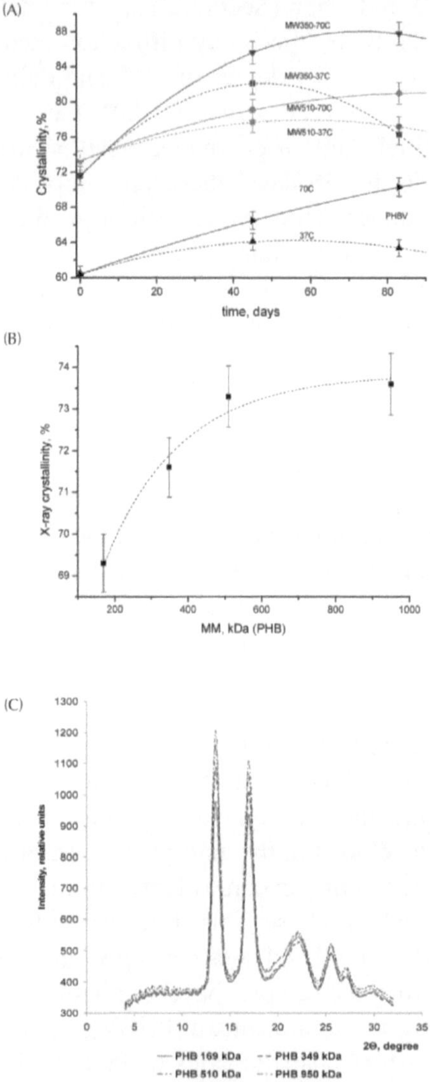

FIGURE 5.3 (a) Crystallinity evolution during the hydrolysis for PHB and PHBV films (denoted values of temperature and MW). (b) Crystallinity as function of initial MW for PHB films prepared by cast method. (c) X-ray diffractograms for PHB films with different molecular weight given under *x*-axis.

Thus, at physiological temperature the crystallinity, measured during degradation by XRD technique has an slightly extreme character. On the initial stage of PHB degradation the crystalline/amorphous ratio is increased owing to additional crystallization through involvement of polymer molecules situated in amorphous fields. In contrast, at 70 °C after reaching the critical MW values (see Section 2), the following desorption of water-soluble intermediates occurs. On the following stage, as the degradation is developed till film disintegration, the crystallinity drop must takes place as result of crystallite disruption.

5.3.4 THE ANALYSIS OF FILM SURFACES FOR PHB BY AFM TECHNIQUE

Morphology and surface roughness of PHB film exposed to corrosive medium (phosphate buffer) have been studied by the AFM technique. This experiment is important for surface characterization because the state of implant surface determines not only mechanism of degradation but the protein adsorption and cell adhesion, which are responsible for polymer biocompatibility [30]. As the standard sample we have used the PHB film with relatively low MW=170 kDa. The film casting procedure may lead to distinction in morphology between two surfaces when the one plane of the polymer film was adjacent with glass plate and the other one was exposed to air. Really, as it is shown in Fig. 5.4 the surface exposed to air has a roughness formed by a plenty of pores with the length of 500–700 nm. The opposite side of the film contacted with glass (Fig.5.4b) is characterized by minor texture and by the pores with the less length as small as 100 nm. At higher magnification (here not presented) in certain localities it can see the stacks of polymer crystallites with width about 100 nm and length 500–800 nm Fig. 5.4.

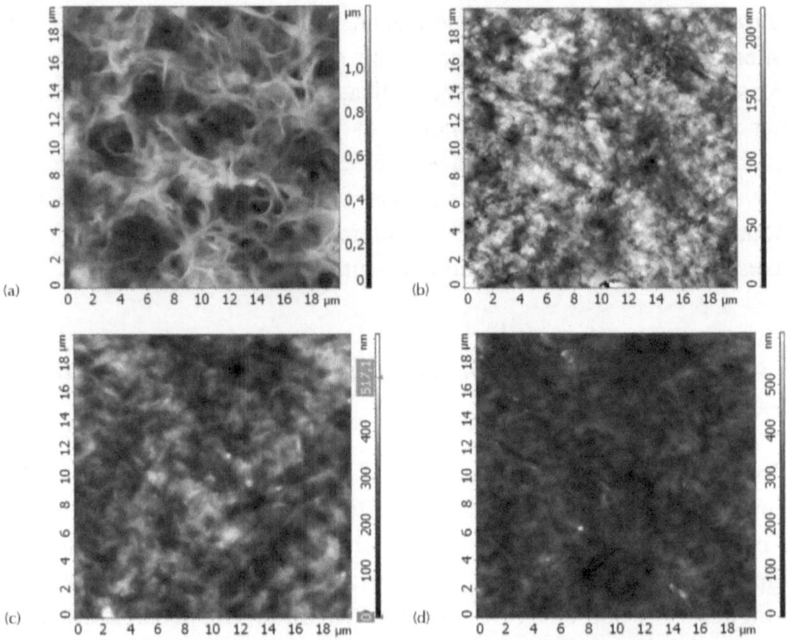

FIGURE 5.4 AFM topographic images of PHB films (170 kDa) with a scan size of 18×18 µm: (a) the rough surface of fresh-prepared sample (exposed to air); (b) the smooth surface of fresh-prepared sample (exposed to glass); (c) the sample exposed to phosphate buffer at 37 °C for 83 days; (d) the sample exposed to phosphate buffer at 70 °C for 83 days. General magnificence is 300.

Inequality of morphology between two surfaces gets clearly evident when quantitative parameters of roughness (r_n) were compared. A roughness analysis has shown that averaged value of this characteristic and a root mean square roughness for surfaces exposed to glass or air differ about ten times.

$$R_{a} = \frac{1}{N}\sum_{n=1}^{N}\left|r_n\right|$$

$$R_q = \sqrt{\frac{1}{N}\sum_{n=1}^{N} r_n^2}$$

The variance of characteristics is related with solvent desorption conditions during its evaporation for the cast film. During chloroform evaporation from the surface faced to air, the flux forms additional channels (viz. the pores), which are fixed as far as the film is solidified and crystallized. Simultaneously, during evaporation the morphology and texture on the opposite side of film exposed to the glass support are not subjected to the impact of solvent transport. The morphology of the latter surface depends on energy interaction conditions (interface glass-biopolymer tension) predominantly. The exposition of PHB films to the buffer for long time (83 days) lead to a threefold growth of roughness characteristics for glass-exposed surface and practically does not affects the air-exposed surface. It is interesting that temperature of film degradation does not influence on the roughness change. The surface characteristics of film surface have the same values after treatment at 37 °C and 70 °C.

Summarizing the AMF data we can conclude that during degradation the air-exposed, rough surface remained stable that probably related with the volume mechanism of degradation (V-mechanism [31, 32]). The pores on the surface provide the fast water diffusion into the bulk of PHB. However, under the same environmental conditions, the change of surface porosity (roughness) for glass-exposed surface is remarkable showing the engagement of surface into degradation process (S-mechanism [31, 32]). Last findings show that along with the volume processes of polymer degradation the surface hydrolysis can proceed. Several authors [20, 21] have recently reported on surface mechanism of PHB destruction but traditional point of view states a volume mechanism of degradation [12]. Here, using an advanced method of surface investigation (AMF) we have shown that for the same film under the same exterior conditions the mechanism of degradation could be changed depending on the prehistory of polymer preparation.

5.4 CONCLUSION

Analyzing all results related with hydrolytic degradation of PHB and its derivatives, the consecutive stages of such complicated process are presented as follows. During the initial stage, the total weight is invariable and the cleavage of bio-molecules resulting in the MW decrease is observed. Within this time the PHB intermediates are too large and hydrophobic to provide solubility in aqueous media. Because the PHB crystallites stay stable, the crystallinity degree is constant as well and even it may grow up due to additional crystallization. On the second stage of hydrolysis, when the MW of intermediates attain the "critical" value, which is equal about 30 kDa, these intermediates can dissolve and diffuse from the polymer into buffer. Within this period the weight loss is clearly observed. The intensity of hydrolysis characterized by the weight loss and the MW decrement is enhanced in the series PHBV < PHB < PHB-PLA blend < PLA.

The growth of initial MW (a terminal group reducing) impacts on the hydrolysis stability probably due to the increase of crystallite perfection and crystallinity degree. The XRD data reflect this trend (see Fig. 5.3b). Moreover, the surface state of PHB films explored by AFM technique depends on the condition of film preparation. After cast processing, there is a great difference in morphologies of PHB film surfaces exposed to air and to glass plate. It is well known that the mechanism of hydrolysis could include two consecutive processes: (i) volume degradation; and (ii) surface degradation. Under essential pore formation (in the surface layer exposed to air) the volume mechanism prevails. The smooth surface of PHB film contacted during preparation with the glass plate is degraded much intensely than the opposite rough surface (Fig. 5.4).

In conclusion, we have revealed that the biopolymer MW determines the form of a hydrolysis profile (see Fig. 5.1). For acceleration of this process we have to use the small MW values of PHB. In this case we affect both the degradation rate and the crystalline degree (Fig 5.3b). By contrast, for prolongation of service-time in a living system it is preferable to use the high-MW PHB that is the most stable polymer against hydrolytic degradation.

KEYWORDS

- **Biopolymer**
- **Crystalline degree**
- **Degradation rate**
- **Hydrolysis profile**
- **Hydrolytic degradation**
- **Polyblend**

REFERENCES

1. Sudesh, K., Abe, H., & Doi, Y. (2000). Synthesis, Structure and Properties of Polyhy-Droxyalkanoates: Biological Polyesters. *Progress in Polymer Science (Oxford), 25(10),* 1503–1555.
2. Lenz, R. W., & Marchessault, R. H. (2005). Bacterial Polyesters: Biosynthesis, Biodegradable Plastics and Biotechnology. *Biomacromolecules, 6(1),* 1–8.
3. Bonartsev, A. P., Iordanskii, A. L., Bonartseva, G. A., & Zaikov, G. E. (2008). Biodegradation and Medical Application of Microbial Poly (3-Hydroxybutyrate). *Polymers Research Journal, 2(2),* 127–160
4. Kadouri, D., Jurkevitch, E., Okon, Y., & Castro-Sowinski, S. (2005). Critical Reviews in Microbiology *31(2),* 55–67 Ecological and Agricultural Significance of Bacterial Polyhydroxyalkanoates.
5. Jendrossek, D., & Handrick, R. (2002). Microbial Degradation of Polyhydroxyalkanoates. *Annu Rev Microbiol, 56,* 403–432.
6. Steinbuchel, A., & Lutke-Eversloh, T. (2003). Metabolic Engineering and Pathway Construction for Biotechnological Production of Relevant Poly Hydroxyalkanoates in Microorganisms. *Biochem. Eng. J., 16,* 81–96.
7. Miller, N. D., & Williams, D. F. (1987). On the Biodegradation of Poly Beta-Hydroxybutyrate (PHB) Homopolymer and Poly Beta-Hydroxybutyrate-Hydroxyvalerate Copolymers. *Biomaterials, 8(2),* 129–137.
8. Qu, X. H., Wu, Q., Zhang, K. Y., & Chen, G. Q. (2006). In Vivo Studies of Poly (3-hydroxybutyrate-co3-hydroxyhexanoate) Based Polymers: Biodegradation and Tissue Reactions. *Biomaterials, 27(19),* 3540–3548
9. Fostera, L. J. R., Sanguanchaipaiwonga, V., Gabelisha, C. L., Hookc, J., & Stenzel, M. (2005). A Natural-Synthetic Hybrid Copolymer of Polyhydroxyoctanoate-Diethylene Glycol: Biosynthesis and Properties. *Polymer, 46,* 6587–6594
10. Marois, Y., Zhang, Z., Vert, M., Deng, X., Lenz, R., & Guidoin, R. (2000). Mechanism and Rate of Degradation of Polyhydroxyoctanoate Films in Aqueous Media: A Long-Term in Vitro Study. *Journal of Biomedical Materials Research, 49(2),* 216–224.
11. Freier, T., Kunze, C., Nischan, C., Kramer, S., Sternberg, K., Sass, M., Hopt, U. T., & Schmitz, K-P. (2002). In Vitro and in Vivo Degradation Studies for Development of a Biodegradable Patch based on Poly(3-hydroxybutyrate). *Biomaterials, 23,* 2649–2657.
12. Doi, Y., Kanesawa, Y., Kawaguchi, Y., & Kunioka, M. (1989). Hydrolytic Degradation of Microbial Poly(hydroxyalkanoates). *Makrom Chem Rapid Commun, 10,* 227–230.
13. Renstadt, R., Karlsson, S., & Albertsson, A. C. (1998). The Influence of Processing Conditions on the Properties and the Degradation of Poly(3-hydroxybutyrate-co3-hydroxyvalerate). *Macromol Symp, 127,* 241–249.
14. Cheng, Mei-Ling, Chen, Po-Ya., Lan Chin-Hung, Sun Yi-Ming. (2011). Structure, Mechanical Properties and Degradation Behaviors of the Electrospun Fibrous Blends of PHBHHx/PDLLA. *Polymer, 52,* 6587–6594.
15. Myshkina, V. L., Nikolaeva, D. A., Makhina, T. K., Bonartsev, A. P., & Bonartseva, G. A. (2008). Effect of Growth Conditions on the Molecular Weight of Poly 3-Hy-

droxybutyrate Produced by Azotobacter Chroococcum 7B. *Applied Biochemistry and Microbiology, 44(5),* 482–486.
16. Myshkina, V. L., Ivanov, E. A., Nikolaeva, D. A., Makhina, T. K., Bonartsev, A. P., Filatova, E. V., Ruzhitsky, A. O., & Bonartseva, G. A. (2010). Biosynthesis of Poly 3-Hydroxybutyrate–3-Hydroxyvalerate Copolymer by Azotobacter Chroococcum Strain 7B. *Applied Biochemistry and Microbiology, 46(3),* 289–296.
17. Akita, S., Einaga, Y., Miyaki, Y., & Fujita, H. (1976). Solution Properties of Poly(D-β-hydroxybutyrate). 1. Biosynthesis and Characterization. *Macromolecules,* 9, 774–780.
18. Rebrov, A. V., Dubinskii, V. A., Nekrasov, Y. P., Bonartseva, G. A., Shtamm, M., Antipov, E. M. (2002). Structure Phenomena at Elastic Deformation of Highly Oriented Polyhydroxy butyrate. *Polymer Science (Russian), 44A,* 347–351.
19. Koyama, N., & Doi, Y. (1995). Morphology and Biodegradability of a Binary Blend of Poly((R)-3-Hydroxybutyric Acid) and Poly((R, S)-lactic acid). *Can. J. Microbiol, 41(Suppl. 1),* 316–322.
20. Majid, M. I. A., Ismail, J., Few, L. L., & Tan, C. F. (2002). The Degradation Kinetics of Poly(3-Hydroxybutyrate) Under Non-Aqueous and Aqueous Conditions. *European Polymer Journal, 38(4),* 837–839.
21. Choi, G. G., Kim, H. W., & Rhee, Y. H. (2004). Enzymatic and Non-Enzymatic Degradation of Poly(3-Hydroxybutyrate-Co-3-Hydroxyvalerate) Copolyesters Produced by Alcaligenes Sp. MT-16. *The Journal of Microbiology, 42(4),* 346–352.
22. Iordanskii, A. L., Rudakova, T. E., & Zaikov, G. E. (1984). Interaction of Polymers with Corrosive and Bioactive Media, VSP, New York -Tokyo
23. Wang, H. T., Palmer, H., Linhardt, R. J., Flanagan, D. R., & Schmitt, E. (1990). Degradation of Poly(ester) Microspheres. *Biomaterials, 11(9),* 679–685.
24. Kurcok, P., Kowalczuk, M., Adamus, G., Jedlinrski, Z., & Lenz, R. W. (1995). Degradability of Poly (b-Hydroxybutyrate)s. Correlation with Chemical Microstucture. *JMS-Pure Appl. Chem, A32*, 875–880.
25. Reusch, R. N. (1992). Biological Complexes of Poly β-Hydroxybutyrate. FEMS *Microbiol. Rev., 103,* 119–130.
26. Molnár, K., Móczó, J. Murariu, M., Dubois, Ph., & Pukánszky, B. (2009) Factors Affecting the Properties of PLA/CaSO4 Composites: Homogeneity and Interactions. *eXPRESS Polymer Letters 3(1),* 49–61
27. Spyros, A., Kimmich, R., Briese, B., & Jendrossek, D. (1997). 1H NMR Imaging Study of Enzymatic Degradation in Poly(3-hydroxybutyrate) and Poly(3-hydroxybutyrate-co-3-hydroxyvalerate). Evidence for Preferential Degradation of Amorphous Phase by PHB Depolymerase B From Pseudomonas Lemoignei. *Macromolecules, 30,* 8218–8225.
28. Luizier, W. D. (1992). Materials Derived from Biomass/Biodegradable Materials. *Proc. Natl. Acad. Sci. USA, 89,* 839–842.
29. Gao, Y., Kong, L., Zhang, L., Gong, Y., Chen, G., Zhao, N., et al. (2006). *Eur Polym J, 42(4),* 764–75.
30. Pompe, T., Keller, K., Mothes, G., Nitschke, M., & Teese, M., Zimmermann, R., Werner, C. (2007). Surface Modification of Poly(hydroxybutyrate) Films to Control Cell-Matrix Adhesion. *Biomaterials, 28(1),* 28–37.

31. Siepmann, J., Siepmann, F., & Florence, A. T. (2006). Local Controlled Drug Delivery to the Brain: Mathematical Modeling of the Underlying Mass Transport Mechanisms. *International Journal of Pharmaceutics, 314(2),* 101–119.
32. Zhang, T. C., Fu, Y. C., Bishop, P. L., et al. (1995). Transport and Biodegradation of Toxic Organics in Biofilms. *Journal of Hazardous Materials, 41(2–3),* 267–285.

CHAPTER 6

POLYMERIC NANOCOMPOSITES REINFORCEMENT

G. V. KOZLOV, YU. G. YANOVSKII, and G. E. ZAIKOV

CONTENTS

ABSTRACT

In this chapter, different methods of filler structure (distribution) determination in polymer matrix; both experimental and theoretical are discussed in detail.

6.1 INTRODUCTION

The experimental analysis of particulate-filled nanocomposites butadiene-styrene rubber/fullerene-containing mineral (nanoshungite) was fulfilled with the aid of force-atomic microscopy, nanoindentation methods and computer treatment. The theoretical analysis was carried out within the frameworks of fractal analysis. It has been shown that interfacial regions in the mentioned nanocomposites are the same reinforcing element as nanofiller actually. The conditions of the transition from nano- to microsystems were discussed. The fractal analysis of nanoshungite particles aggregation in polymer matrix was performed. In has been shown that reinforcement of the studied nanocomposites is a true nanoeffect.

The modern methods of experimental and theoretical analysis of polymer materials structure and properties allow not only to confirm earlier propounded hypotheses, but to obtain principally new results. Let us consider some important problems of particulate-filled polymer nanocomposites, the solution of which allows to advance substantially in these materials properties understanding and prediction. Polymer nanocomposites multicomponentness (multiphaseness) requires their structural components quantitative characteristics determination. In this aspect interfacial regions play a particular role, since it has been shown earlier, that they are the same reinforcing element in elastomeric nanocomposites as nanofiller actually [1]. Therefore, the knowledge of interfacial layer dimensional characteristics is necessary for quantitative determination of one of the most important parameters of polymer composites in general their reinforcement degree [2, 3].

The aggregation of the initial nanofiller powder particles in more or less large particles aggregates always occurs in the course of technological process of making particulate filled polymer composites in general [4] and elastomeric nanocomposites in particular [5]. The aggregation process tells on composites (nanocomposites) macroscopic properties [2–4]. For nanocomposites nanofiller aggregation process gains special significance, since its intensity can be the one, that nanofiller particles aggregates size exceeds 100 nm the value, which is assumed (though conditionally enough [6]) as an upper dimensional limit for nanoparticle. In other words, the aggregation process can result to the situation when primordially supposed nanocomposite ceases to be one. Therefore, at present several methods exist, which allow to suppress nanoparticles aggregation process [5, 7]. This also assumes the necessity of the nanoparticles aggregation process quantitative analysis.

It is well known [1, 2], that in particulate-filled elastomeric nanocomposites (rubbers) nanofiller particles form linear spatial structures ("chains"). At the same time in polymer composites, filled with disperse microparticles (microcomposites) particles (aggregates of particles) of filler form a fractal network, which defines polymer matrix structure (analog of fractal lattice in computer simulation) [4]. This results to different mechanisms of polymer matrix structure formation in micro and nanocomposites. If in the first filler particles (aggregates of particles) fractal network availability results to "disturbance" of polymer matrix structure, that is expressed in the increase of its fractal dimension d_f [4], then in case of polymer nanocomposites at nanofiller contents change the value d_f is not changed and equal to matrix polymer structure fractal dimension [3]. As it has been expected, the change of the composites of the indicated classes structure formation mechanism change defines their properties, in particular, reinforcement degree [11, 12]. Therefore, nanofiller structure fractality strict proof and its dimension determination are necessary.

As it is known [13, 14], the scale effects in general are often found at different materials mechanical properties study. The dependence of failure stress on grain size for metals (Holl-Petsch formula) [15] or of effective filling degree on filler particles size in case of polymer composites [16] are examples of such effect. The strong dependence of elasticity modulus on nanofiller particles diameter is observed for particulate-filled elastomeric nanocomposites [5]. Therefore, it is necessary to elucidate the physical grounds of nano- and micromechanical behavior scale effect for polymer nanocomposites.

At present a disperse material wide list is known, which is able to strengthen elastomeric polymer materials [5]. These materials are very diverse on their surface chemical constitution, but particles small size is a common feature for them. On the basis of this observation the hypothesis was offered, that any solid material would strengthen the rubber at the condition that it was in a very dispersed state and it could be dispersed in polymer matrix. Edwards [5] points out, that filler particles small size is necessary and, probably, the main requirement for reinforcement effect realization in rubbers. Using modern terminology, one can say, that for rubbers reinforcement the nanofiller particles, for which their aggregation process is suppressed as far as possible, would be the most effective ones [3, 12]. Therefore, the theoretical analysis of a nanofiller particles size influence on polymer nanocomposites reinforcement is necessary.

Proceeding from the said above, this chapter purpose is the solution of the considered above paramount problems with the help of modern experimental and theoretical techniques on the example of particulate-filled butadiene-styrene rubber.

6.2 EXPERIMENTAL PART

The made industrially butadiene-styrene rubber of mark SKS-30, which contains 7.0–12.3% cis- and 71.8–72.0% trans-bonds, with density of 920–930 kg/m^3 was used as matrix polymer. This rubber is fully amorphous one.

Fullerene-containing mineral shungite of Zazhoginsk's deposit consists of ~30% globular amorphous metastable carbon and ~70% high-disperse silicate particles. Besides, industrially made technical carbon of mark № 220 was used as nanofiller. The technical carbon, nano- and microshugite particles average size makes up 20, 40 and 200 nm, respectively. The indicated filler content is equal to 37 mass %. Nano- and micro dimensional disperse shungite particles were prepared from industrially output material by the original technology processing. The size and polydispersity analysis of the received in milling process shungite particles was monitored with the aid of analytical disk centrifuge (CPS Instruments, Inc., USA), allowing to determine with high precision size and distribution by the sizes within the range from 2 nm up to 50 mcm.

Nanostructure was studied on atomic-forced microscopes Nano-DST (Pacific Nanotechnology, USA) and Easy Scan DFM (Nanosurf, Switzerland) by semicontact method in the force modulation regime. Atomic-force microscopy results were processed with the help of specialized software package SPIP (Scanning Probe Image Processor, Denmark). SPIP is a powerful programs package for processing of images, obtained on SPM, AFM, STM, scanning electron microscopes, transmission electron microscopes, interferometers, confocal microscopes, profilometers, optical microscopes and so on. The given package possesses the whole functions number, which are necessary at images precise analysis, in a number of which the following ones are included:

- the possibility of three-dimensional reflecting objects obtaining, distortions automatized leveling, including Z-error mistakes removal for examination of separate elements and so on;
- quantitative analysis of particles or grains, more than 40 parameters can be calculated for each found particle or pore: area, perimeter, mean diameter, the ratio of linear sizes of grain width to its height distance between grains, coordinates of grain center of mass a.a. can be presented in a diagram form or in a histogram form.

The tests on elastomeric nanocomposites nanomechanical properties were carried out by a nano indentation method [17] on apparatus Nano Test 600 (Micro Materials, Great Britain) in loads wide range from 0.01 mN up to 2.0 mN. Sample indentation was conducted in 10 points with interval of 30 mcm. The load was increased with constant rate up to the greatest given load reaching (for the rate 0.05 mN/s-1 mN). The indentation rate was changed in conformity with the greatest load value counting, that loading cycle should take 20 s. The unloading was conducted with the same rate as loading. In the given experiment the "Berkovich indentor" was used with the angle at the top of 65.3° and rounding radius of 200 nm. Indentations were carried out in the checked load regime with preload of 0.001 mN.

For elasticity modulus calculation the obtained in the experiment by nanoindentation course dependences of load on indentation depth (strain) in ten points for each sample at loads of 0.01, 0.02, 0.03, 0.05, 0.10, 0.50, 1.0 and 2.0 mN were processed according to Oliver-Pharr method [18].

6.3 RESULTS AND DISCUSSION

In Fig. 6.1, the obtained according to the original methodics results of elasticity moduli calculation for nanocomposite butadiene-styrene rubber/nanoshungite components (matrix, nanofiller particle and interfacial layers), received in interpolation process of nanoindentation data, are presented. The processed in SPIP polymer nanocomposite image with shungite nanoparticles allows experimental determination of interfacial layer thickness l_{if}, which is presented in Fig. 6.1 as steps on elastomeric matrix-nanofiller boundary. The measurements of 34 such steps (interfacial layers) width on the processed in SPIP images of interfacial layer various section gave the mean experimental value l_{if}=8.7 nm. Besides, nanoindentation results (Fig. 6.1, figures on the right) showed, that interfacial layers elasticity modulus was only by 23–45% lower than nanofiller elasticity modulus, but it was higher than the corresponding parameter of polymer matrix in 6.0–8.5 times. These experimental data confirm, that for the studied nanocomposite interfacial layer is a reinforcing element to the same extent, as nanofiller actually [1, 3, 12].

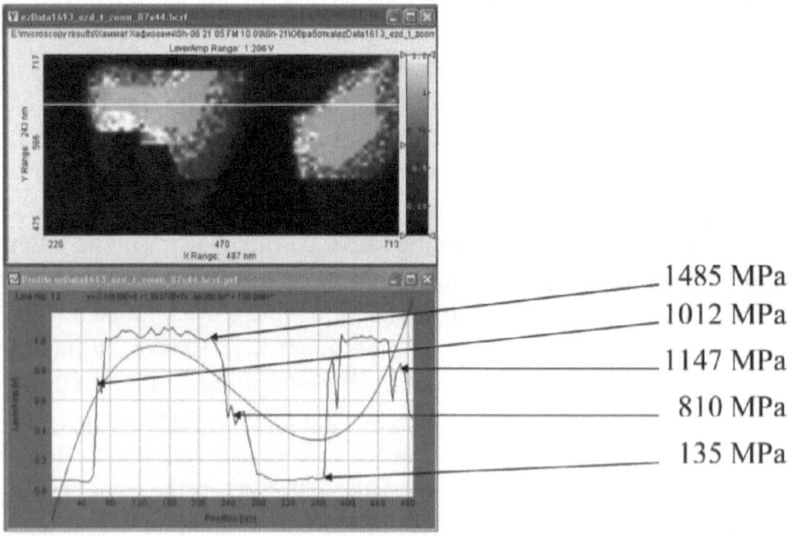

FIGURE 6.1 The processed in SPIP image of nanocomposite butadiene-styrene rubber/nanoshungite, obtained by force modulation method, and mechanical characteristics of structural components according to the data of nano-indentation (strain 150 nm).

Let us fulfill further the value l_{if} theoretical estimation according to the two methods and compare these results with the ones obtained experimentally. The first method simulates interfacial layer in polymer composites as a result of interaction of two fractals polymer matrix and nanofiller surface [19, 20]. In this case there is a sole linear scale l, which defines these fractals interpenetration distance [21]. Since nanofiller elasticity modulus is essentially higher, than the corresponding parameter for rubber (in the considered case in 11 times, see Fig. 6.1), then the indicated interaction reduces to nanofiller indentation in polymer matrix and then $l=l_{if}$. In this case it can be written [21]:

$$l_{if} \approx a\left(\frac{R_p}{a}\right)^{2(d-d_{surf})/d}, \tag{1}$$

where a is a lower linear scale of fractal behavior, which is accepted for polymers as equal to statistical segment length l_{st} [22], R_p is a nanofiller particle (more precisely, particles aggregates) radius, which for nanoshungite is equal to ~ 84 nm [23], d is dimension of Euclidean space, in which fractal is considered (it is obvious, that in our case d=3), d_{surf} is fractal dimension of nanofiller particles aggregate surface.

The value l_{st} is determined as follows [24]:

$$l_{st} = l_0 C_\infty, \tag{2}$$

where l_0 is the main chain skeletal bond length, which is equal to 0.154 nm for both blocks of butadiene-styrene rubber [25], C is characteristic ratio, which is a polymer chain statistical flexibility indicator [26], and is determined with the help of the equation [22]:

$$T_g = 129\left(\frac{S}{C_\infty}\right)^{1/2}, \tag{3}$$

where T_g is glass transition temperature, equal to 217 K for butadiene-styrene rubber [3], S is macromolecule cross-sectional area, determined for the mentioned rubber according to the additivity rule from the following considerations. As it is known [27], the macromolecule diameter quadrate values are equal: for polybutadiene 20.7 Å^2 and for polystyrene 69.8 Å^2. Having calculated cross-sectional area of macromolecule, simulated as a cylinder, for the indicated polymers according to the known geometrical formulas, let us obtain 16.2 and 54.8 Å^2, respectively. Further, accepting as S the average value of the adduced above areas, let us obtain for butadiene-styrene rubber S=35.5 Å^2. Then according to the Eq. (3) at the indicated values T_g and S let us obtain C=12.5 and according to the Eq. (2) – l_{st}=1.932 nm.

The fractal dimension of nanofiller surface d_{surf} was determined with the help of the equation [3]:

$$S_u = 410 R_p^{d_{surf} - d}, \tag{4}$$

where S_u is nanoshungite particles specific surface, calculated as follows [28]:

$$S_u = \frac{3}{\rho_n R_p}, \tag{5}$$

where ρ_n is the nanofiller particles aggregate density, determined according to the formula [3]:

$$\rho_n = 0.188 \left(R_p \right)^{1/3}. \tag{6}$$

The calculation according to the Eqs. (4)–(6) gives d_{surf}=2.44. Further, using the calculated by the indicated mode parameters, let us obtain from the equation (1) the theoretical value of interfacial layer thickness l_{if}^T=7.8 nm. This value is close enough to the obtained one experimentally (their discrepancy makes up ~ 10%).

The second method of value l_{if}^T estimation consists in using of the two following equations [3, 29]:

$$\varphi_{if} = \varphi_n \left(d_{surf} - 2 \right) \tag{7}$$

and

$$\varphi_{if} = \varphi_n \left[\left(\frac{R_p + l_{if}^T}{R_p} \right)^3 - 1 \right], \tag{8}$$

where φ_{if} and φ_n are relative volume fractions of interfacial regions and nanofiller, accordingly.

The combination of the indicated equations allows to receive the following formula for l_{if}^T calculation:

$$l_{if}^T = R_p \left[\left(d_{surf} - 1 \right)^{1/3} - 1 \right]. \tag{9}$$

The calculation according to the Eq. (9) gives for the considered nanocomposite l_{if}^T=10.8 nm, that also corresponds well enough to the experiment (in this case discrepancy between l_{if} and l_{if}^T makes up ~ 19%).

Let us note in conclusion the important experimental observation, which follows from the processed by program SPIP results of the studied nanocomposite surface scan (Fig. 6.1). As one can see, at one nanoshungite particle surface from one to three (in average two) steps can be observed, structurally identified as interfacial layers. It is significant that these steps width (or l_{if}) is approximately equal to the first (the closest to nanoparticle surface) step width. Therefore, the indicated observation supposes, that in elastomeric nanocomposites at average two interfacial layers are formed: the first at the expense of nanofiller particle surface with elastomeric matrix interaction, as a result of which molecular mobility in this layer is frozen and its state is glassy-like one, and the second at the expense of glassy interfacial layer with elastomeric polymer matrix interaction. The most important question from the practical point of view, whether one interfacial layer or both serve as nanocomposite reinforcing element. Let us fulfill the following quantitative estimation for this question solution. The reinforcement degree (E_n/E_m) of polymer nanocomposites is given by the equation [3]:

$$\frac{E_n}{E_m} = 1 + 11\left(\varphi_n + \varphi_{if}\right)^{1.7}, \tag{10}$$

where E_n and E_m are elasticity moduli of nanocomposite and matrix polymer, accordingly (E_m=1.82 MPa [3]).

According to the Eq. (7) the sum (φ_n+φ_{if}) is equal to:

$$\varphi_n + \varphi_{if} = \varphi_n\left(d_{surf} - 1\right), \tag{11}$$

If one interfacial layers (the closest to nanoshungite surface) is a reinforcing element and if both interfacial layers are a reinforcing element.

$$\varphi_n + 2\varphi_{if} = \varphi_n\left(2d_{surf} - 3\right), \tag{12}$$

In its turn, the value φ_n is determined according to the equation [30]:

$$\varphi_n = \frac{W_n}{\rho_n}, \tag{13}$$

where W_n is nanofiller mass content, ρ_n is its density, determined according to the Eq. (6).

The calculation according to the Eqs. (11) and (12) gave the following E_n/E_m values: 4.60 and 6.65, respectively. Since the experimental value E_n/E_m=6.10 is closer to the value, calculated according to the Eq. (12), then this means that both interfacial layers are a reinforcing element for the studied nanocomposites. Therefore, the coefficient 2 should be introduced in the equations for value l_{if} determination (e.g., in the Eq. (1)) in case of nanocomposites with elastomeric matrix. Let us remind that the Eq. (1) in its initial form was obtained as a relationship with proportionality sign, that is, without fixed proportionality coefficient [21].

Thus, the used above nanoscopic methodics allow to estimate both interfacial layer structural special features in polymer nanocomposites and its sizes and properties. For the first time it has been shown, that in elastomeric particulate-filled nanocomposites two consecutive interfacial layers are formed, which are a reinforcing element for the indicated nanocomposites. The proposed theoretical methodics of interfacial layer thickness estimation, elaborated within the frameworks of fractal analysis, give well enough correspondence to the experiment.

For theoretical treatment of nanofiller particles aggregate growth processes and final sizes traditional irreversible aggregation models are inapplicable, since it is obvious, that in nanocomposites aggregates a large number of simultaneous growth takes place. Therefore, the model of multiple growths, offered in Ref. [6], was used for nanofiller aggregation description.

In Fig. 6.2, the images of the studied nanocomposites, obtained in the force modulation regime, and corresponding to them nanoparticles aggregates fractal dimension d_f distributions are adduced. As it follows from the adduced values d_f^{ag} (d_f^{ag}=2.40–2.48), nanofiller particles aggregates in the studied nanocomposites are formed by a mechanism particle-cluster (P-Cl), that is, they are Witten-Sander clusters [32]. The variant A, was chosen which according to mobile particles are added to the lattice, consisting of a large number of "seeds" with density of c_0 at simulation beginning [31]. Such model generates the structures, which have fractal geometry on length short scales with value d_f≈2.5 (see Fig. 6.2) and homogeneous structure on length large scales. A relatively high particles concentration c is required in the model for uninterrupted network formation [31].

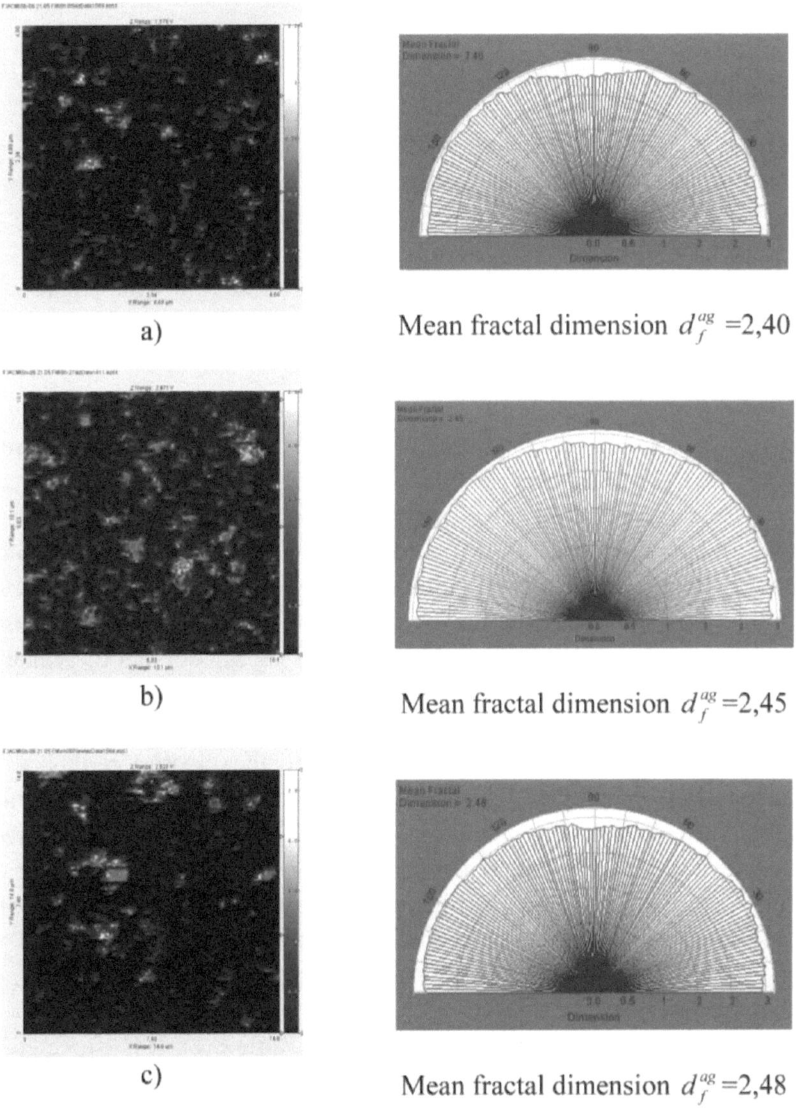

FIGURE 6.2 The images, obtained in the force modulation regime, for nanocomposites, filled with technical carbon (a), nanoshungite (b), microshungite (c) and corresponding to them fractal dimensions d_f^{ag} .

In case of "seeds" high concentration c_0 for the variant A the following relationship was obtained [31]:

$$R_{max}^{d_f^{ag}} = N = c / c_0 , \tag{14}$$

where R_{max} is nanoparticles cluster (aggregate) greatest radius, N is nanoparticles number per one aggregate, c is nanoparticles concentration, c_0 is "seeds" number, which is equal to nanoparticles clusters (aggregates) number.

The value N can be estimated according to the following equation [8]:

$$2R_{max} = \left(\frac{S_n N}{\pi \eta} \right)^{1/2} , \tag{15}$$

where S_n is cross-sectional area of nanoparticles, of which an aggregate consists, η is a packing coefficient, equal to 0.74 [28].

The experimentally obtained nanoparticles aggregate diameter $2R_{ag}$ was accepted as $2R_{max}$ (Table 6.1) and the value S_n was also calculated according to the experimental values of nanoparticles radius r_n (Table 6.1). In Table 6.1 the values N for the studied nanofillers, obtained according to the indicated method, were adduced. It is significant that the value N is a maximum one for nanoshungite despite larger values r_n in comparison with technical carbon.

Further the Eq. (14) allows to estimate the greatest radius R_{max}^T of nanoparticles aggregate within the frameworks of the aggregation model [31]. These values R_{max}^T are adduced in Table 6.1, from which their reduction in a sequence of technical carbon-nanoshungite-microshungite, that fully contradicts to the experimental data, that is, to R_{ag} change (Table 6.1). However, we must not neglect the fact that the Eq. (14) was obtained within the frameworks of computer simulation, where the initial aggregating particle sizes are the same in all cases [31]. For real nanocomposites the values r_n can be distinguished essentially (Table 6.1). It is expected, that the value R_{ag} or R_{max}^T will be the higher, the larger is the radius of nanoparticles, forming aggregate, is, that is, r_n. Then theoretical value of nanofiller particles cluster (aggregate) radius R_{ag}^T can be determined as follows:

$$R_{ag}^T = k_n r_n N^{1/d_f^{ag}} , \tag{16}$$

where k_n is proportionality coefficient, in the present work accepted empirically equal to 0.9.

TABLE 6.1 The Parameters of Irreversible Aggregation Model of Nanofiller Particles Aggregates Growth

Nanofiller	R_{ag}, nm	r_n, nm	N	R^T_{max}, nm	R^T_{ag}, nm	R_c, nm
Technical carbon	34.6	10	35.4	34.7	34.7	33.9
Nanoshungite	83.6	20	51.8	45.0	90.0	71.0
Microshungite	117.1	100	4.1	15.8	158.0	255.0

The comparison of experimental R_{ag} and calculated according to the equation (16) R^T_{ag} values of the studied nanofillers particles aggregates radius shows their good correspondence (the average discrepancy of R_{ag} and R^T_{ag} makes up 11.4 %). Therefore, the theoretical model [31] gives a good correspondence to the experiment only in case of consideration of aggregating particles real characteristics and, in the first place, their size.

Let us consider two more important aspects of nanofiller particles aggregation within the frameworks of the model [31]. Some features of the indicated process are defined by nanoparticles diffusion at nanocomposites processing. Specifically, length scale, connected with diffusible nanoparticle, is correlation length ξ of diffusion. By definition, the growth phenomena in sites, remote more than ξ, are statistically independent. Such definition allows to connect the value ξ with the mean distance between nanofiller particles aggregates L_n. The value ξ can be calculated according to the equation [31]:

$$\xi^2 \approx \tilde{n}^{-1} R_{ag}^{d_f^{ag}-d+2}, \qquad (17)$$

where c is nanoparticles concentration, which should be accepted equal to nanofiller volume contents φ_n, which is calculated according to the Eqs. (6) and (13).

The values r_n and R_{ag} were obtained experimentally (see histogram of Fig. 6.3). In Fig. 6.4 the relation between L_n and ξ is adduced, which, as it is expected, proves to be linear and passing through coordinates origin. This means, that the distance between nanofiller particles aggregates is limited by mean displacement of statistical walks, by which nanoparticles are simulated. The relationship between L_n and ξ can be expressed analytically as follows:

$$L_n \approx 9.6\xi, \text{ nm.} \tag{18}$$

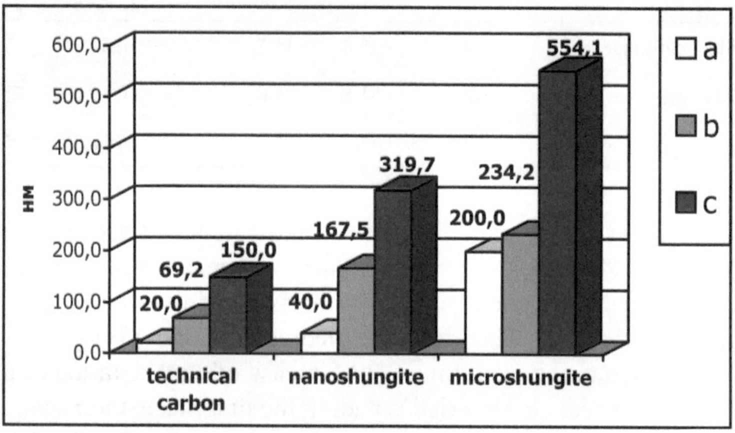

FIGURE 6.3 The initial particles diameter (a), their aggregates size in nanocomposite (b) and distance between nanoparticles aggregates (c) for nanocomposites, filled with technical carbon, nano- and microshungite.

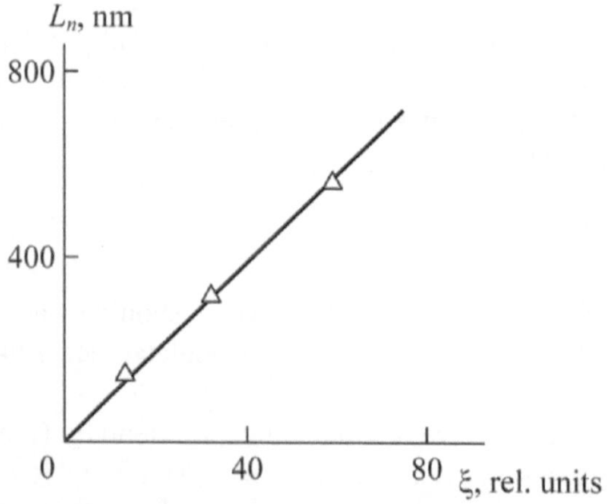

FIGURE 6.4 The relation between diffusion correlation length ξ and distance between nanoparticles aggregates L_n for considered nanocomposites.

The second important aspect of the model [31] in reference to nanofiller particles aggregation simulation is a finite nonzero initial particles concentration c or φ_n effect, which takes place in any real systems. This effect is realized at the condition $\xi \approx R_{ag}$, that occurs at the critical value $R_{ag}(R_c)$, determined according to the relationship [31]:

$$c \sim R_c^{d_f^{ag}-d}. \tag{19}$$

The Eq. (19) right side represents cluster (particles aggregate) mean density. This equation establishes that fractal growth continues only, until cluster density reduces up to medium density, in which it grows. The calculated according to the Eq. (19) values R_c for the considered nanoparticles are adduced in Table 6.1, from which follows, that they give reasonable correspondence with this parameter experimental values R_{ag} (the average discrepancy of R_c and R_{ag} makes up 24%).

Since the treatment [31] was obtained within the frameworks of a more general model of diffusion-limited aggregation, then its correspondence to the experimental data indicated unequivocally, that aggregation processes in these systems were controlled by diffusion. Therefore, let us consider briefly nanofiller particles diffusion. Statistical walkers diffusion constant ζ can be determined with the aid of the relationship [31]:

$$\xi \approx (\zeta t)^{1/2}, \tag{20}$$

where t is walk duration.

The Eq. (20) supposes (at t=const) ζ increase in a number technical carbon-nanoshungite-microshungite as 196–1069–3434 relative units, that is, diffusion intensification at diffusible particles size growth. At the same time diffusivity D for these particles can be described by the well-known Einstein's relationship [33]:

$$D = \frac{kT}{6\pi\eta r_n \alpha}, \tag{21}$$

where k is Boltzmann constant, T is temperature, η is medium viscosity, α is numerical coefficient, which further is accepted equal to 1.

In its turn, the value η can be estimated according to the equation [34]:

$$\frac{\eta}{\eta_0} = 1 + \frac{2.5\phi_n}{1-\phi_n}, \tag{22}$$

where η_0 and η are initial polymer and its mixture with nanofiller viscosity, accordingly.

The calculation according to the equations (21) and (22) shows, that within the indicated above nanofillers number the value D changes as 1.32–1.14–0.44 relative units, that is, reduces in three times, that was expected. This apparent contradiction is due to the choice of the condition t=const (where t is nanocomposite production duration) in the Eq. (20). In real conditions the value t is restricted by nanoparticle contact with growing aggregate and then instead of t the value t/c_0 should be used, where c_0 is the seeds concentration, determined according to the Eq. (14). In this case the value ζ for the indicated nanofillers changes as 0.288–0.118–0.086, that is, it reduces in 3.3 times that corresponds fully to the calculation according to the Einstein's relationship (the Eq. (21)). This means, that nanoparticles diffusion in polymer matrix obeys classical laws of Newtonian rheology [33].

Thus, the disperse nanofiller particles aggregation in elastomeric matrix can be described theoretically within the frameworks of a modified model of irreversible aggregation particle-cluster. The obligatory consideration of nanofiller initial particles size is a feature of the indicated model application to real systems description. The indicated particles diffusion in polymer matrix obeys classical laws of Newtonian liquids hydrodynamics. The offered approach allows to predict nanoparticles aggregates final parameters as a function of the initial particles size, their contents and other factors number.

At present there are several methods of filler structure (distribution) determination in polymer matrix, both experimental [10, 35] and theoretical [4]. All the indicated methods describe this distribution by fractal dimension D_n of filler particles network. However, correct determination of any object fractal (Hausdorff) dimension includes three obligatory conditions. The first from them is the indicated above determination of fractal dimension numerical magnitude, which should not be equal to object topological dimension. As it is known [36], any real (physical) fractal possesses fractal properties within a certain scales range. Therefore, the second condition is the evidence of object self-similarity in this scales range [37]. And at last, the third condition is the correct choice of measurement scales range itself. As it has been shown in Refs. [38, 39], the minimum range should exceed at any rate one self-similarity iteration.

The first method of dimension D_n experimental determination uses the following fractal relationship [40, 41]:

$$D_n = \frac{\ln N}{\ln \rho}, \tag{23}$$

where N is a number of particles with size ρ.

Particles sizes were established on the basis of atomic-power microscopy data (see Fig. 6.2). For each from the three studied nanocomposites no less than 200 particles were measured, the sizes of which were united into 10 groups and mean values N and ρ were obtained. The dependences $N(\rho)$ in double logarithmic coordinates were plotted, which proved to be linear and the values D_n were calculated according to their slope (see Fig. 6.5). It is obvious, that at such approach fractal dimension D_n is determined in two-dimensional Euclidean space, whereas real nanocomposite should be considered in three-dimensional Euclidean space. The following relationship can be used for D_n recalculation for the case of three-dimensional space [42]:

$$D3 = \frac{d + D2 \pm \left[(d - D2)^2 - 2\right]^{1/2}}{2}, \tag{24}$$

where $D3$ and $D2$ are corresponding fractal dimensions in three- and two-dimensional Euclidean spaces, d=3.

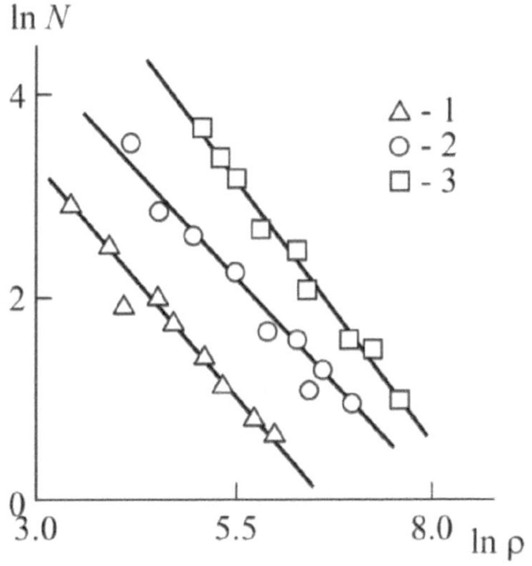

FIGURE 6.5 The dependences of nanofiller particles number N on their size ρ for nanocomposite BSR/TC (1), BSR/nanoshungite (2) and BSR/microshungite (3).

The calculated according to the indicated method dimensions D_n are adduced in Table 6.2. As it follows from the data of this table, the values D_n for the studied nanocomposites are varied within the range of 1.10–1.36, that is, they characterize more or less branched linear formations ("chains") of nanofiller particles (aggregates of particles) in elastomeric nanocomposite structure. Let us remind that for particulate-filled composites polyhydroxiether/graphite the value D_n changes within the range of ~2.30–2.80 [4, 10], that is, for these materials filler particles network is a bulk object, but not a linear one [36].

Another method of D_n experimental determination uses the so-called "quadrates method" [43]. Its essence consists in the following. On the enlarged nanocomposite microphotograph (see Fig. 6.2) a net of quadrates with quadrate side size α_i, changing from 4.5 up to 24 mm with constant ratio α_{i+1}/α_i=1.5, is applied and then quadrates number N_i, in to which nanofiller particles hit (fully or partly), is counted up. Five arbitrary net positions concerning microphotograph were chosen for each measurement. If nanofiller particles network is a fractal, then the following relationship should be fulfilled [43]:

$$N_i \sim S_i^{-D_n/2}, \qquad (25)$$

where S_i is quadrate area, which is equal to α_i^2.

In Fig. 6.6, the dependences of N_i on S_i in double logarithmic coordinates for the three studied nanocomposites, corresponding to the Eq. (25), is adduced. As one can see, these dependences are linear, that allows to determine the value D_n from their slope. The determined according to the Eq. (25) values D_n are also adduced in Table 6.2, from which a good correspondence of dimensions D_n, obtained by the two described above methods, follows (their average discrepancy makes up 2.1% after these dimensions recalculation for three-dimensional space according to the Eq. (24)).

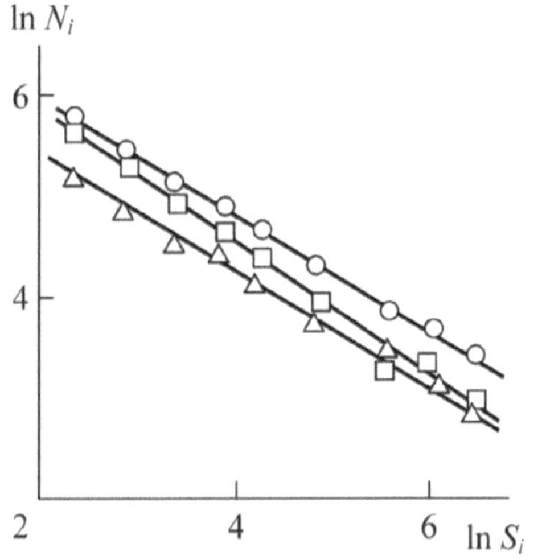

FIGURE 6.6 The dependences of covering quadrates number N_i on their area S_i, corresponding to the Eq. (25), in double logarithmic coordinates for nanocomposites on the basis of BSR. The designations are the same, that in Fig. 6.5.

As it has been shown in Ref. [44], the usage for self-similar fractal objects at the Eq. (25) the condition should be fulfilled:

$$N_i - N_{i-1} \sim S_i^{-D_n} . \tag{26}$$

In Fig. 6.7, the dependence, corresponding to the Eq. (26), for the three studied elastomeric nanocomposites is adduced. As one can see, this dependence is linear, passes through coordinates origin that according to the Eq. (26) is confirmed by nanofiller particles (aggregates of particles) "chains" self-similarity within the selected α_i range. It is obvious, that this self-similarity will be a statistical one [44]. Let us note, that the points, corresponding to α_i=16 mm for nanocomposites butadiene-styrene rubber/technical carbon (BSR/TC) and butadiene-styrene rubber/microshungite (BSR/microshungite), do not correspond to a common straight line. Accounting for electron microphotographs of Fig. 6.2 enlargement this gives the self-similarity range for nanofiller "chains" of 464–1472 nm. For nanocomposite butadiene-styrene rubber/nanoshungite (BSR/nanoshungite), which has no points deviating from a straight line of Fig. 6.7, α_i range makes up 311–1510 nm, that corresponds well enough to the indicated above self-similarity range.

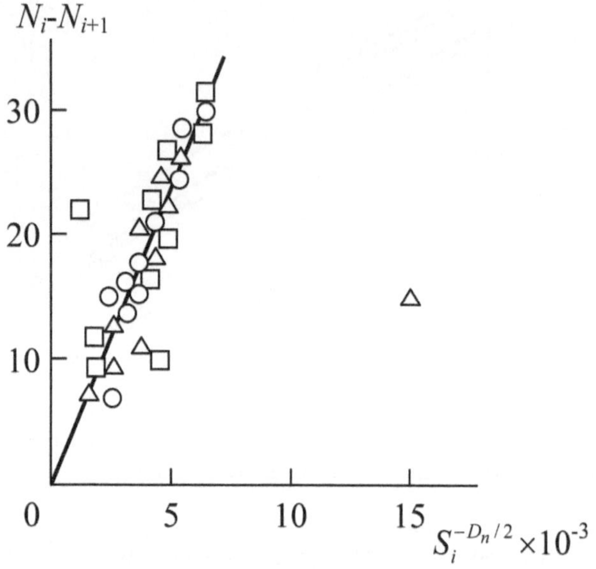

FIGURE 6.7 The dependences of (N_i–N_{i+1}) on the value $S_i^{-D_n/2}$, corresponding to the relationship (26), for nanocomposites on the basis of BSR. The designations are the same, that in Fig. 6.5.

In Refs. [38, 39] it has been shown, that measurement scales S_i minimum range should contain at least one self-similarity iteration. In this case the condition for ratio of maximum S_{max} and minimum S_{min} areas of covering quadrates should be fulfilled [39]:

$$\frac{S_{max}}{S_{min}} > 2^{2/D_n}. \tag{27}$$

Hence, accounting for the defined above restriction let us obtain S_{max}/S_{min}=121/20.25=5. 975, that is larger than values $2^{2/D_n}$ for the studied nanocomposites, which are equal to 2.71–3.52. This means, that measurement scales range is chosen correctly.

The self-similarity iterations number μ can be estimated from the inequality [39]:

$$\left(\frac{S_{max}}{S_{min}}\right)^{D_n/2} > 2^{\mu} . \tag{28}$$

Using the indicated above values of the included in the inequality Eq. (28) parameters, μ=1.42–1.75 is obtained for the studied nanocomposites, that is, in our experiment conditions self-similarity iterations number is larger than unity, that again confirms correctness of the value D_n estimation [35].

And let us consider in conclusion the physical grounds of smaller values D_n for elastomeric nanocomposites in comparison with polymer microcomposites, that is, the causes of nanofiller particles (aggregates of particles) "chains" formation in the first ones. The value D_n can be determined theoretically according to the equation [4]:

$$\phi_{if} = \frac{D_n + 2.55d_0 - 7.10}{4.18}, \tag{29}$$

where φ_{if} is interfacial regions relative fraction, d_0 is nanofiller initial particles surface dimension.

The dimension d_0 estimation can be carried out with the help of the Eq. (4) and the value φ_{if} can be calculated according to the Eq. (7). The results of dimension D_n theoretical calculation according to the Eq. (29) are adduced in (Table 6.2), from which a theory and experiment good correspondence follows. The Eq. (29) indicates unequivocally to the cause of a filler in nano- and microcomposites different behavior. The high (close to 3, see Table 6.2) values d_0 for nanoparticles and relatively small (d_0=2.17 for graphite [4]) values d_0 for microparticles at comparable values φ_{if} is such cause for composites of the indicated classes [3, 4].

TABLE 6.2 The Dimensions of Nanofiller Particles (Aggregates of Particles) Structure in Elastomeric Nanocomposites

Nanocomposite	D_n, the Eq. (23)	D_n, the Eq. (25)	d_0	d_{surf}	φ_n	D_n, the Eq. (29)
BSR/TC	1.19	1.17	2.86	2.64	0.48	1.11
BSR/nanoshungite	1.10	1.10	2.81	2.56	0.36	0.78
BSR/microshungite	1.36	1.39	2.41	2.39	0.32	1.47

Hence, the stated above results have shown, that nanofiller particles (aggregates of particles) "chains" in elastomeric nanocomposites are physical fractal within self-similarity (and, hence, fractality [41]) range of ~500–1450 nm. In this range their dimension D_n can be estimated according to the Eqs. (23), (25) and (29). The cited examples demonstrate the necessity of the measurement scales range correct choice. As it has been noted earlier [45], the linearity of the plots, corresponding to the Eqs. (23) and (25), and D_n nonintegral value do not guarantee object self-similarity (and, hence, fractality). The nanofiller particles (aggregates of particles) structure low dimensions are due to the initial nanofiller particles surface high fractal dimension.

In Fig. 6.8, the histogram is adduced, which shows elasticity modulus E change, obtained in nanoindentation tests, as a function of load on indenter P or nanoindentation depth h. Since for all the three considered nanocomposites the dependences $E(P)$ or $E(h)$ are identical qualitatively, then further the dependence $E(h)$ for nanocomposite BSR/TC was chosen, which reflects the indicated scale effect quantitative aspect in the most clearest way.

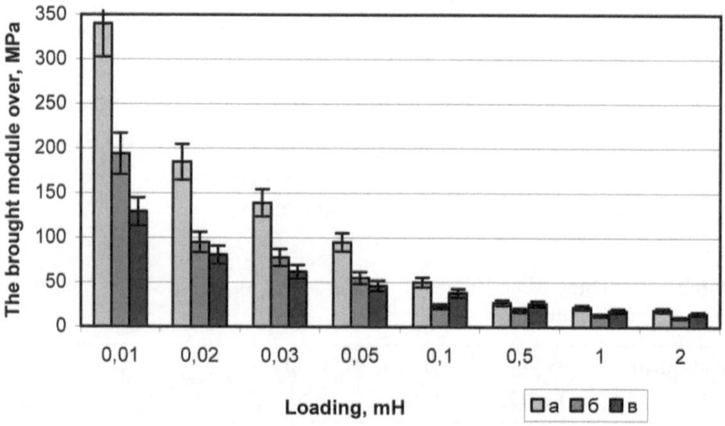

FIGURE 6.8 The dependences of reduced elasticity modulus on load on indentor for nanocomposites on the basis of butadiene-styrene rubber, filled with technical carbon (a), micro (b) and nanoshungite (c).

In Fig. 6.9, the dependence of E on h_{pl} (see Fig. 6.10) is adduced, which breaks down into two linear parts. Such dependences elasticity modulus strains are typical for polymer materials in general and are due to intermolecular bonds anharmonicity [46]. In Ref. [47], it has been shown that the dependence $E(h_{pl})$ first part at $h_{pl} \leq 500$ nm is not connected with relaxation processes and has a purely elastic origin. The elasticity modulus E on this part changes in proportion to h_{pl} as:

$$E = E_0 + B_0 h_{pl}, \tag{30}$$

where E_0 is "initial" modulus, that is, modulus, extrapolated to h_{pl}=0, and the coefficient B_0 is a combination of the first and second kind elastic constants. In the considered case B_0<0. Further Grüneisen parameter γ_L, characterizing intermolecular bonds an harmonicity level, can be determined [47]:

$$\gamma_L \approx -\frac{1}{6} - \frac{1}{2}\frac{B_0}{E_0}\frac{1}{(1-2\nu)}, \tag{31}$$

where ν is Poisson ratio, accepted for elastomeric materials equal to ~ 0.475 [36].

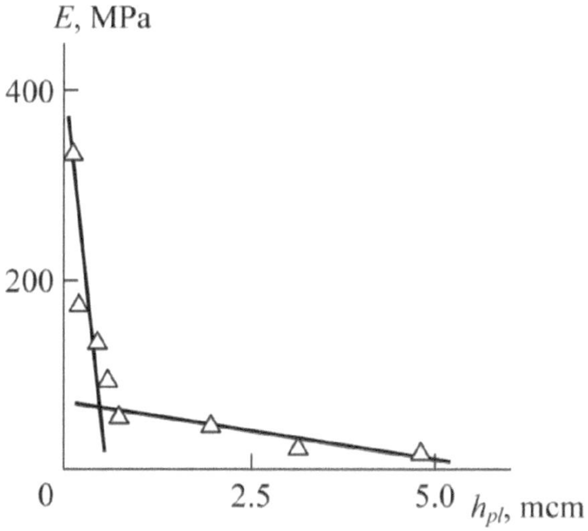

FIGURE 6.9 The dependence of reduced elasticity modulus E obtained in nano indentation experiment on plastic strain h_{pl} for nanocomposites BSR/TC.

Calculation according to the Eq. (31) has given the following values γ_L: 13.6 for the first part and 1.50 for the second one. Let us note the first from γ_L adduced values is typical for intermolecular bonds, whereas the second value γ_L is much closer to the corresponding value of Grüneisen parameter G for intrachain modes [46].

Poisson's ratio ν can be estimated by γ_L (or G) known values according to the formula [46]:

$$\gamma_L = 0.7\left(\frac{1+\nu}{1-2\nu}\right). \tag{32}$$

The estimations according to the Eq. (32) gave: for the dependence $E(h_{pl})$ first part ν=0.462, for the second one – ν=0.216. If for the first part the value ν is close to Poisson's ratio magnitude for nonfilled rubber [36], then in the second part case the additional estimation is required. As it is known [48], a polymer composites (nanocomposites) Poisson's ratio value ν_n can be estimated according to the equation:

$$\frac{1}{\nu_n} = \frac{\phi_n}{\nu_{\text{TC}}} + \frac{1-\phi_n}{\nu_m}, \tag{33}$$

where φ_n is nanofiller volume fraction, ν_{TC} and ν_m are nanofiller (technical carbon) and polymer matrix Poisson's ratio, respectively.

The value ν_m is accepted equal to 0.475 [36] and the magnitude ν_{TC} is estimated as follows [49]. As it is known [50], the nanoparticles TC aggregates fractal dimension d_f^{ag} value is equal to 2.40 and then the value ν_{TC} can be determined according to the equation [50]:

$$d_f^{ag} = (d-1)(1+\nu_{\text{TC}}). \tag{34}$$

According to the Eq. (34) ν_{TC}=0.20 and calculation ν_n according to the Eq. (33) gives the value 0.283, that is close enough to the value ν=0.216 according to the Eq. (32) estimation. The obtained by the indicated methods values ν and ν_n comparison demonstrates, that in the dependence $E(h_{pl})$ (h_{pl}<0.5 mcm) the first part in nanoindentation tests only rubber-like polymer matrix ($\nu=\nu_m\approx0.475$) is included and in this dependence the second part–the entire nanocomposite as homogeneous system [51] – $\nu=\nu_n\approx0.22$.

Let us consider further E reduction at h_{pl} growth (Fig. 6.9) within the frameworks of density fluctuation theory, which value ψ can be estimated as follows [22]:

$$\psi = \frac{\rho_n kT}{K_T}, \tag{35}$$

where ρ_n is nanocomposite density, k is Boltzmann constant, T is testing temperature, K_T is isothermal modulus of dilatation, connected with Young's modulus E by the relationship [46]:

$$K_T = \frac{E}{3(1-\nu)} \tag{36}$$

In Fig. 6.10, the scheme of volume of the deformed at nanoindentation material V_{def} calculation in case of Berkovich indentor using is adduced and in Fig. 6.11, the dependence $\psi(V_{def})$ in logarithmic coordinates was shown. As it follows from the data of this figure, the density fluctuation growth is observed at the deformed material volume increase. The plot $\psi(\ln V_{def})$ extrapolation to ψ=0 gives $\ln V_{def}\approx 13$ or $V_{def}(V_{def}^{cr})=4.42\times10^5$ nm^3. Having determined the linear scale l_{cr} of transition to ψ=0 as $(V_{def}^{cr})^{1/3}$, let us obtain l_{cr}=75.9 nm, that is close to nanosystems dimensional range upper boundary (as it was noted above, conditional enough [6]), which is equal to 100 nm. Thus, the stated above results suppose, that nanosystems are such systems, in which density fluctuations are absent, always taking place in microsystems.

As it follows from the data of Fig. 6.9, the transition from nano- to microsystems occurs within the range h_{pl}=408–726 nm. Both the indicated above values h_{pl} and the corresponding to them values $(V_{def})^{1/3}\approx$814–1440 nm can be chosen as the linear length scale l_n, corresponding to this transition. From the comparison of these values l_n with the distance between nanofiller particles aggregates L_n (L_n=219.2–788.3 nm for the considered nanocomposites, (see Fig. 6.3) it follows, that for transition from nano- to microsystems l_n should include at least two nanofiller particles aggregates and surrounding them layers of polymer matrix, that is the lowest linear scale of nanocomposite simulation as a homogeneous system. It is easy to see, that nanocomposite structure homogeneity condition is harder than the obtained above from the criterion ψ=0. Let us note, that such method, namely, a nanofiller particle and surrounding it polymer matrix layers separation, is widespread at a relationships derivation in microcomposite models.

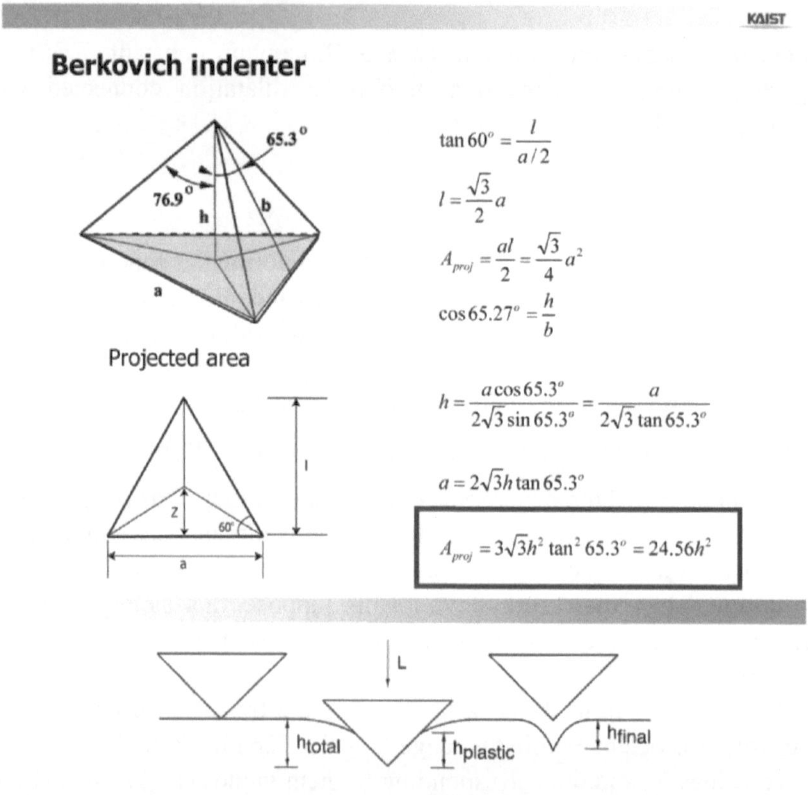

FIGURE 6.10 The schematic images of Berkovich indentor and nanoindentation process.

It is obvious, that the Eq. (35) is inapplicable to nanosystems, since $\psi \rightarrow 0$ assumes $K_T \rightarrow \infty$ that is physically incorrect. Therefore, the value E_0, obtained by the dependence $E(h_{pl})$ extrapolation (see Fig. 6.9) to $h_{pl}=0$, should be accepted as E for nanosystems [49].

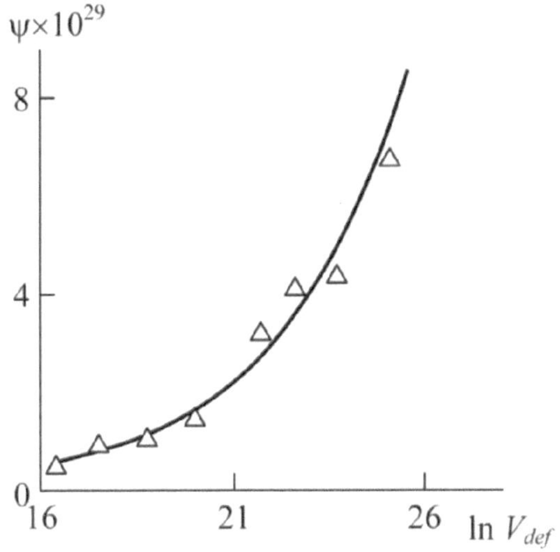

FIGURE 6.11 The dependence of density fluctuation ψ on volume of deformed in nanoindentation process material V_{def} in logarithmic coordinates for nanocomposites BSR/TC.

Hence, the stated above results have shown, that elasticity modulus change at nanoindentation for particulate-filled elastomeric nanocomposites is due to a number of causes, which can be elucidated within the frameworks of an harmonicity conception and density fluctuation theory. Application of the first from the indicated conceptions assumes, that in nanocomposites during nano indentation process local strain is realized, affecting polymer matrix only, and the transition to macrosystems means nanocomposite deformation as homogeneous system. The second from the mentioned conceptions has shown, that nano- and micro systems differ by density fluctuation absence in the first and availability of ones in the second. The last circumstance assumes that for the considered nanocomposites density fluctuations take into account nanofiller and polymer matrix density difference. The transition from nano to Microsystems is realized in the case, when the deformed material volume exceeds nanofiller particles aggregate and surrounding it layers of polymer matrix combined volume [49].

In Ref. [3], the following formula was offered for elastomeric nanocomposites reinforcement degree E_n/E_m description:

$$\frac{E_n}{E_m} = 15.2\left[1-\left(d-d_{surf}\right)^{1/t}\right], \tag{37}$$

where t is index percolation, equal to 1.7 [28].

From the Eq. (37) it follows, that nanofiller particles (aggregates of particles) surface dimension d_{surf} is the parameter, controlling nanocomposites reinforcement degree [53]. This postulate corresponds to the known principle about numerous division surfaces decisive role in nanomaterials as the basis of their properties change [54]. From the Eqs. (4)–(6) it follows unequivocally, that the value d_{surf} is defined by nanofiller particles (aggregates of particles) size R_p only. In its turn, from the Eq. (37) it follows, that elastomeric nanocomposites reinforcement degree E_n/E_m is defined by the dimension d_{surf} only, or, accounting for the said above, by the size R_p only. This means, that the reinforcement effect is controlled by nanofiller particles (aggregates of particles) sizes only and in virtue of this is the true nanoeffect.

In Fig. 6.12, the dependence of E_n/E_m on $(d\text{-}d_{surf})^{1/1.7}$ is adduced, corresponding to the equation (37), for nanocomposites with different elastomeric matrices (natural and butadiene-styrene rubbers, NR and BSR, accordingly) and different nanofillers (technical carbon of different marks, nano- and microshungite). Despite the indicated distinctions in composition, all adduced data are described well by the Eq. (37).

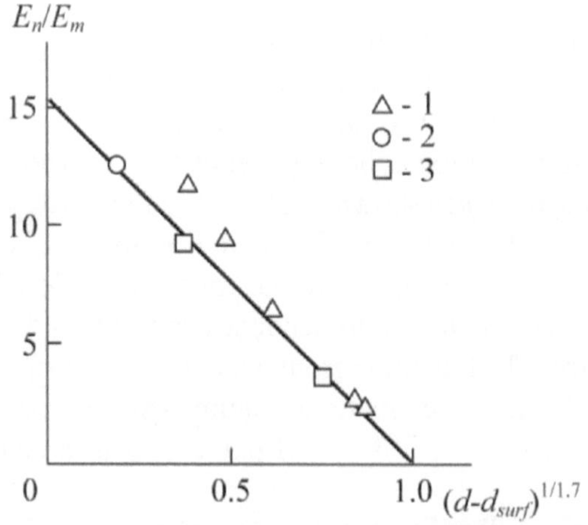

FIGURE 6.12 The dependence of reinforcement degree E_n/E_m on parameter $(d\text{-}d_{surf})^{1/1.7}$ value for nanocomposites NR/TC (1), BSR/TC (2) and BSR/shungite (3).

In Fig. 6.13, two theoretical dependences of E_n/E_m on nanofiller particles size (diameter D_p), calculated according to the Eqs. (4)–(6) and (37), are adduced. However, at the curve 1 calculation the value D_p for the initial nanofiller particles was used and at the curve 2 calculation – nanofiller particles aggregates size D_p^{ag} (see Fig. 6.3). As it was expected [5], the growth E_n/E_m at D_p or D_p^{ag} reduction, in addition the calculation with D_p (nonaggregated nanofiller) using gives higher E_n/E_m values in comparison with the aggregated one (D_p^{ag} using). At $D_p \leq 50$ nm faster growth E_n/E_m at D_p reduction is observed than at $D_p > 50$ nm, that was also expected. In Fig. 6.13, the critical theoretical value D_p^{cr} for this transition, calculated according to the indicated above general principles [54], is pointed out by a vertical shaded line. In conformity with these principles the nanoparticles size in nanocomposite is determined according to the condition, when division surface fraction in the entire nanomaterial volume makes up about 50% and more. This fraction is estimated approximately by the ratio $3l_{if}/D_p$, where l_{if} is interfacial layer thickness. As it was noted above, the data of (Fig. 6.1) gave the average experimental value $l_{if} \approx 8.7$ nm. Further from the condition $3l_{if}/D_p \approx 0.5$ let us obtain $D_p \approx 52$ nm that is shown in Fig. 6.13 by a vertical shaded line. As it was expected, the value $D_p \approx 52$ nm is a boundary one for regions of slow ($D_p > 52$ nm) and fast ($D_p \leq 52$ nm) E_n/E_m growth at D_p reduction. In other words, the materials with nanofiller particles size $D_p \leq 52$ nm ("super reinforcing" filler according to the terminology of Ref. [5]) should be considered true nanocomposites.

Let us note in conclusion, that although the curves 1 and 2 of Fig. 6.13 are similar ones, nanofiller particles aggregation, which the curve 2 accounts for, reduces essentially enough nanocomposites reinforcement degree. At the same time the experimental data correspond exactly to the curve 2 that was to be expected in virtue of aggregation processes, which always took place in real composites [4] (nanocomposites [55]). The values d_{surf}, obtained according to the Eqs. (4)–(6), correspond well to the determined experimentally ones. So, for nanoshungite and two marks of technical carbon the calculation by the indicated method gives the following d_{surf} values: 2.81, 2.78 and 2.73, whereas experimental values of this parameter are equal to: 2.81, 2.77 and 2.73, that is, practically a full correspondence of theory and experiment was obtained.

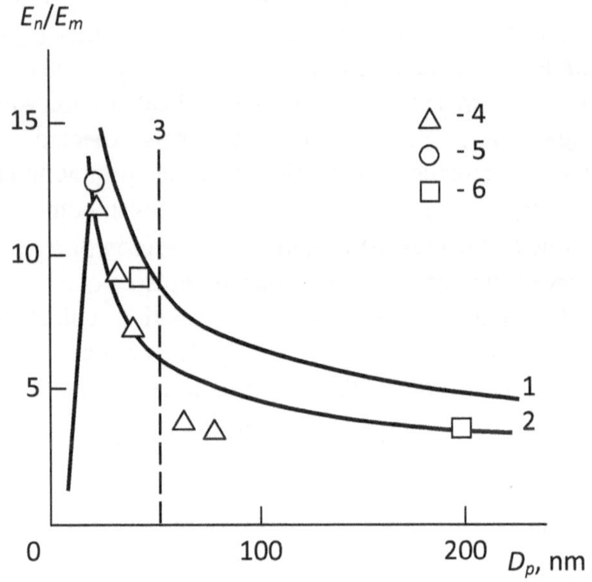

FIGURE 6.13 The theoretical dependences of reinforcement degree E_n/E_m on nanofiller particles size D_p, calculated according to the Eqs. (4)–(6) and (37), at initial nanoparticles (1) and nanoparticles aggregates (2) size using. 3 – the boundary value D_p, corresponding to true nanocomposite. 4–6 – the experimental data for nanocomposites NR/TC (4), BSR/TC (5) and BSR/shungite (6).

6.4 CONCLUSIONS

Hence, the stated above results have shown, that the elastomeric reinforcement effect is the true nanoeffect, which is defined by the initial nanofiller particles size only. The indicated particles aggregation, always taking place in real materials, changes reinforcement degree quantitatively only, namely, reduces it. This effect theoretical treatment can be received within the frameworks of fractal analysis. For the considered nanocomposites the nanoparticle size upper limiting value makes up ~ 52 nm.

KEYWORDS

- **Nanocomposites**
- **Nanofillers**
- **Polymer reinforcement**

REFERENCES

1. Yanovskii, Yu. G., Kozlov, G. V., & Karnet, Yu. N. (2011). *Mekhanika Kompozitsionnykh Materialov i Konstruktsii, 17(2),* 203–208.
2. Malamatov, A. Kh., Kozlov, G. V., & Mikitaev, M. A. (2006). *Reinforcement Mechanisms of Polymer Nanocomposites.* Moscow, Publishers of the D.I. Mendeleev RKhTU, 240 p.
3. Mikitaev, A. K., Kozlov, G. V., & Zaikov, G. E. (2009). *Polymer Nanocomposites: Variety of Structural Forms and Applications.* Moscow, Nauka, 278 p.
4. Kozlov, G. V., Yanovskii, Yu. G., & Karnet, Yu. N. (2008). *Structure and Properties of Particulate-Filled Polymer Composites: the Fractal Analysis.* Moscow, Al'yanstransatom, 363 p.
5. Edwards, D. C. (1990). J. *Mater. Sci., 25(12),* 4175–4185.
6. Buchachenko, A. L. (2003). *Uspekhi Khimii, 72(5),* 419–437.
7. Kozlov, G. V., Yanovskii, Yu. G., Burya, A. I., & Aphashagova, Z. Kh. (2007). *Mekhanika Kompozitsionnykh Materialov i Konstruktsii, 13(4),* 479–492.
8. Lipatov, Yu. S. (1977). *The Physical Chemistry of Filled Polymers.* Moscow, Khimiya, 304 p.
9. Bartenev, G. M., & Zelenev, Yu. V. (1983). *Physics and Mechanics of Polymers.* Moscow, Vysshaya Shkola, 391 p.
10. Kozlov, G. V., & Mikitaev, A. K. (1996). *Mekhanika Kompozitsionnykh Materialov i Konstruktsii, 2(3–4),* 144–157.
11. Kozlov, G. V., Yanovskii, Yu. G., & Zaikov, G. E. (2010). *Structure and Properties of Particulate-Filled Polymer Composites*: the Fractal Analysis. New York, Nova Science Publishers, Inc, 282 p.
12. Mikitaev, A. K., Kozlov, G. V., & Zaikov, G. E. (2008). *Polymer Nanocomposites*: Variety of Structural Forms and Applications. New York, Nova Science Publishers, Inc, 319 p.
13. McClintok, F. A., & Argon, A. S. (1966). *Mechanical Behavior of Materials.* Reading, Addison-Wesley Publishing Company, Inc, 440 p.
14. Kozlov, G. V., & Mikitaev, A. K. (1987). *Doklady AN SSSR, 294(5),* 1129–1131.
15. Honeycombe, R. W. K. (1968). *The Plastic Deformation of Metals.* Boston, Edward Arnold (Publishers), Ltd, 398 p.
16. Dickie, R. (1980). A. In book: *Polymer Blends.* New York, San-Francisco, London, Academic Press, *1,* 386–431.

17. Kornev, Yu. V., Yumashev, O. B., Zhogin, V. A., Karnet, Yu. N., & Yanovskii, Yu. G. (2008). *Kautschuk, i Rezina, 6,* 18–23.
18. Oliver, W. C., & Pharr, G. M. (1992). *J. Mater. Res, 7(6),* 1564–1583.
19. Kozlov, G. V., Yanovskii, Yu. G., & Lipatov, Yu. S. (2002). *Mekhanika Kompozitsionnykh Materialov i Konstruktsii, 8(1),* 111–149.
20. Kozlov, G. V., Burya, A. I., & Lipatov, Yu. S. (2006). *Mekhanika Kompozitnykh Materialov, 42(6),* 797–802.
21. Hentschel, H. G. E., & Deutch, J. M. (1984). *Phys. Rev. A, 29(3),* 1609–1611.
22. Kozlov, G. V., Ovcharenko, E. N., & Mikitaev, A. K. (2009). Structure of Polymers Amorphous State. Moscow, Publishers of the D.I. *Mendeleev RKhTU*, 392 p.
23. Yanovskii, Yu. G., & Kozlov, G. V. (2011). Mater. VII Intern. Sci.-Pract. Conf. "New Polymer Composite Materials.*" Nal'chik, KBSU*, 189–194.
24. Wu, S. (1989). *J. Polymer Sci.: Part B: Polymer Phys, 27(4),* 723–741.
25. Aharoni, S. M. (1983). *Macromolecules, 16(9),* 1722–1728.
26. Budtov, V. P. (1992). *The Physical Chemistry of Polymer Solutions*. Sankt-Peterburg, Khimiya, 384 p.
27. Aharoni, S. M. (1985). *Macromolecules, 18(12),* 2624–2630.
28. Bobryshev, A. N., Kozomazov, V. N., Babin, L. O., & Solomatov, V. I. (1994). Synergetics of Composite Materials. *Lipetsk, NPO ORIUS*, 154 p.
29. Kozlov, G. V., Yanovskii, Yu. G., & Karnet, Yu. N. (2005). Mekhanika Kompozitsionnykh *Materialov i Konstruktsii, 11(3),* 446–456.
30. Sheng, N., Boyce, M. C., Parks, D. M., Rutledge, G. C., Abes, J. I., & Cohen, R. E. (2004). *Polymer, 45(2),* 487–506.
31. Witten, T. A., & Meakin, P. (1983). *Phys. Rev. B, 28(10),* 5632–5642.
32. Witten, T. A., & Sander, L. M. (1983). *Phys. Rev. B, 27(9),* 5686–5697.
33. Happel, J., & Brenner, G. (1976). Hydrodynamics at Small Reynolds Numbers. Moscow, *Mir*, 418 p.
34. Mills, N. J. (1971). *J. Appl. Polymer Sci., 15(11),* 2791–2805.
35. Kozlov, G. V., Yanovskii, Yu. G., & Mikitaev, A. K. (1998). *Mekhanika Kompozitnykh Materialov, 34(4),* 539–544.
36. Balankin, A. S. (1991). Synergetics of Deformable Body. Moscow, *Publishers of Ministry Defence SSSR*, 404 p.
37. Hornbogen, E. (1989). *Intern. Mater. Res, 34(6),* 277–296.
38. Pfeifer, P. (1984). *Appl. Surf. Sci., 18(1),* 146–164.
39. Avnir, D., Farin, D., & Pfeifer, P. (1985). *J. Colloid Interface Sci., 103(1),* 112–123.
40. Ishikawa, K. (1990). *J. Mater. Sci. Lett, 9(4),* 400–402.
41. Ivanova, V. S., Balankin, A. S., Bunin, I. Zh., & Oksogoev, A. A. (1994). *Synergetics and Fractals in Material Science*. Moscow, Nauka, 383 p.
42. Vstovskii, G. V., Kolmakov, L. G., & Terent'ev, V. E. (1993). *Metally, 4,* 164–178.
43. Hansen, J. P., & Skjeitorp, A. T. (1988). *Phys. Rev. B, 38(4),* 2635–2638.
44. Pfeifer, P., Avnir, D., & Farin, D. (1984). *J. Stat. Phys, 36(5/6),* 699–716.
45. Farin, D., Peleg, S., Yavin, D., & Avnir, D. (1985). *Langmuir, 1(4),* 399–407.
46. Kozlov, G. V., & Sanditov, D. S. (1994). *Anharmonical Effects and Physical-Mechanical Properties of Polymers.* Novosibirsk, Nauka, 261 p.
47. Bessonov, M. I., & Rudakov, A. P. (1971). Vysokomolek. *Soed.* B, V. 13, *7,* 509–511.

48. Kubat, J., Rigdahl, M., & Welander, M. (1990). *J. Appl. Polymer Sci., 39(5),* 1527–1539.
49. Yanovskii, Yu. G., Kozlov, G. V., Kornev, Yu. V., Boiko, O. V., & Karnet, Yu. N. (2010). *Mekhanika Kompozitsionnykh Materialov i Konstruktsii, 16(3),* 445–453.
50. Yanovskii, Yu. G., Kozlov, G. V., & Aloev, V. Z. (2011). Mater. Intern. Sci.-Pract. Conf. "Modern Problems of APK Innovation Development Theory and Practice." *Nal'chik, KBSSKhA,* 434–437.
51. Chow, T. S. (1991). *Polymer, 32(1),* 29–33.
52. Ahmed, S., & Jones, F. R. (1990). *J. Mater. Sci., 25(12),* 4933–4942.
53 Kozlov, G. V., Yanovskii, Yu. G., & Aloev, V. Z. (2011). Mater. Intern. Sci.-Pract. Conf, dedicated to FMEP 50-th Anniversary. *Nal'chik, KBSSKhA,* 83–89.
54. Andrievskii, R. A. (2002). *Rossiiskii Khimicheskii Zhurnal, 46(5),* 50–56.
55. Kozlov, G. V., Sultonov, N. Zh., Shoranova, L. O., & Mikitaev, A. K. (2011). *Naukoemkie Tekhnologii, 12(3),* 17–22.

CHAPTER 7

AROMATIC POLYESTERS

ZINAIDA S. KHASBULATOVA and GENNADY E. ZAIKOV

CONTENTS

ABSTRACT

The data on aromatic polyesters based on phthalic and *n*-oxybenzoic acid derivatives have been presented and various methods of synthesis of such polyesters developed by scientists from different countries for last 50 years have been reviewed.

7.1 INTRODUCTION

The important trend of modern chemistry and technology of polymeric materials is the search for the possibilities of producing materials with novel properties based on given combination of known polymers.

One of the most interesting ways in this direction is the creation of block-copolymers macromolecules of which are the "hybrids" of units differing in chemical structure and composition. Thermodynamical incompatibility of blocks results in stable microphase layering in the majority cases what, finally, allows one to combine the properties of various fragments of macromolecules of block-copolymers in an original way.

Depending on diversity of chemical nature of blocks, their length, number and sequence as well as their ability to crystallize one can obtain materials of structure and properties distinguishing from that of initial components. Here are the huge potential possibilities the practical realization of which has already been started. The most evident is the creation of thermoplastic elastomers (TPEs) – high-tonnage polymeric materials synthesized on the base of principle of block-copolymerization: joining of properties of both thermoplastics and elastomers in one material. The great potentials of block-copolymers caused considerable attention to them within the last years.

Nowadays, all the main problems of physics and physic-chemistry of polymers became closely intertwined when studying block-copolymers: the nature of ordering in polymers, the features of phase separation in polymers and the influence of general molecular parameters on it, the stability of phases at exposing to temperature and power impacts, the features of physical and mechanical properties of microphases and the role of their conjugation.

Existing today numerous methods of synthesis of block-copolymers give the possibility to combine unlimited number of various macromolecules, what already allowed people to synthesize multiple block-copolymers. The thermal and mechanical properties and also the stability of industrial block-copolymers vary in broad limits.

The range of operating temperatures and the thermal stability of TPEs have lately been extended owing to the use of solid blocks of high T_{glass} (of polysulfones or polycarbonates for instance) combined with soft blocks of low T_{glass}. Moreover, the incompatibility of those blocks results in independence of elasticity modulus on temperature in broad temperature range. Applying appropriate selection of chemical nature of blocks one can also improve the other properties.

Some limitations and unresolved issues still exist in areas of synthesis, analysis and characterization of properties and usage of block-copolymers. This is a good stimulus for the intense researches and development of corresponding fields of industry.

The most preferred methods of synthesis of block-copolymers are the three following. The polymerization according to the mechanism of "live" chains with the consecutive addition of monomers has been used in the first one. The second is based on the interaction of two preliminary obtained oligomers with the end functional groups. The third one is the polycondensation of the second block at expense of end group of primarily obtained block of the first monomer. The second and the third methods allow one to use great variety of chemical structures.

So, to produce block-copolymers one can avail numerous reactions allowing one to bind, within the macromolecule, blocks synthesized by means of polycondensation at expense of joining or cycloreversion.

The second method of synthesis of block-copolymers allows one to produce polymers after various combinations of initial compounds, one among which is that monomers able to enter the reaction of condensation are added to oligomers obtained by polycondensation.

Generally, the bifunctional components are used for creating the block-copolymers of $(–AB–)_n$ type. The necessary oligomers could be obtained by means of either condensation reactions or usual polymerization. The end groups of monomer taken in excess are responsible for the chemical nature of process in case of condensation.

Until now, the morphological studies have been performed mainly on block-copolymers containing two chemically different blocks A and B only. One can expect the revelation of quite novel morphological structures for three-block copolymers, including three mutually incompatible units $(ABC)_n$. And there are few references on such polymers.

Russian and foreign scientists remarkably succeeded in both areas of creation of new inflammable, heat- and thermal resistant polycondensation polymers and areas of development of methods of performing polycondensation and studying of the mechanism of reactions grounding the polycondensation processes [1–6].

The reactions of polycondensation are the bases of producing the most important classes of heterochain polymers: polyarylates, polysulfones, polyarylenesterketones, polycarbonates, polyamides and others [7–12].

Non-equilibrium polycondensation can be characterized by a number of advantages among the polycondensation processes. These are the absence of exchange destructive processes, high values of constants of speed of growth of polymeric chain, et cetera. However, some questions of nonequilibrium polycondensation still remain unanswered: the mechanism and basic laws of formation of copolymers when the possibility of combination of positive properties of two, three or more initial monomers can be realized in high-molecular product.

The simple and complex aromatic polyesters, polysulfones and polyarylene ketones possess the complex of valuable properties such as high physic-mechanical and dielectric properties as well as increased thermal stability.

There is a lot of foreign scientific papers devoted to the synthesis and study of copolysulfones based on oligosulfones and polyarylenesterketones based on oligoketones.

Because of importance of the problem of creation of thermo-stable polymers possessing high flame and thermal resistance accompanied with the good physic-mechanical properties, the study of the regularities of formation of copolyesters and block-copolyesters based of oligosulfoneketones, oligosulfones, oligoketones and oligoformals appeared to be promising depending on the constitution of initial compounds, establishing of interrelations between composition, structure and properties of copolymers.

To improve the basic physic-mechanical parameters and abilities to be reused (in particular, to be dissolved), the synthesis of copolyesters and block-copolyesters has been performed through the stage of formation of oligomers with end reaction-able functional groups.

As the result of performed activities, the oligomers of various chemical compositions have been synthesized: oligosulfones, oligoketones, oligosulfoneketones, oligoformals, and novel aromatic copolyesters and block-copolyesters have been produced.

Obtained copoly and block-copolyestersulfoneketones, as well as polyarylates based of dichloranhydrides of phthalic acids and chloranhydride of 3,5-dibromine-*n*-oxybenzoic acid and copolyester with groups of terephthaloyl-bis(*n*-oxybenzoic) acid possess high mechanical and dielectric properties, thermal and fire resistance and also the chemical stability. The regularities of acceptor-catalyst method of polycondensation and high-temperature polycondensation when synthesizing named polymers have been studied and the relations between the composition, structure and properties of polymers obtained have been established. The synthesized here block-copolyesters and copolyesters can find application in various fields of modern industry (automobile, radioelectronic, electrotechnique, avia, electronic, chemical and others) as thermal resistant construction and layered (film) materials.

7.2 AROMATIC POLYESTERS

Aromatic polyesters are polycondensation organic compounds containing complex ester groups, simple ester links and aromatic fragments within their macromolecule in different combination.

Aromatic polyesters (AP) are thermo-stable polymers; they are thermoplastic products useful for reprocessing into the articles and materials by means of formation methods from solutions and melts.

Mainly, plastics and films are produced from aromatic copolyesters. APs can be used also as lacquers, fibrous binding agents for synthetic paper, membranes, hollow fibers, as additions for semiproducts when obtaining materials based on other polymers.

Many articles based on APs appear in industry. The world production of APs increased from 38 millions of tons in 2004 to 50 in 2008, or on 32%. The polymers referred to the class of constructional plastics are distinguished among APs.

Until now, the technical progress in many fields of industry, especially in engineering industry, instrument production, was circumfused namely by the use of constructional plastics. Such exploitation properties of polymers as durability, thermal stability, electroisolation, antifriction properties, optical transparency and others determine their usage instead of ferrous and nonferrous metals, alloys, wood, ceramics and glass [13]. 1 ton of polymers replace 5–6 tons of ferrous and nonferrous metals and 3–3.5

tons of wood while the economy of labor expenses reaches 800 man-hours per 1 ton of polymers. About 50% of all polymers used in engineering industry are consumed in electrotechnique and electronics. The 80% of all the production of electrotechnique and up to 95% of that of engineering industry has been produced with help of polymers.

The use of construction plastics allows one to create principally new technology of creation the details, machine knots and devices what provides for the high economical efficiency. The construction polymers are well used by means of modern methods: casting and extrusion to the articles operating in conditions of sign-altering loads at temperatures 100–200 °C.

The modern chemical industry gave constructional thermoplastic materials with lowered consumption of material and weight of the machines, devices, mechanisms, reduced power capacities and labor-intensity when manufacturing and exploiting, increased stint.

Nowadays, the radioelectronics, electrotechnical, avia, shipbuilding and fields of industry can not develop successively without using the modern progressive polymers such as polyarylates, polysulfones, polyesterketones and others, which are perspective construction materials. Only Russian industry involves 50 types of plastics including more than 850 labels and various modifications [14]. As a result, the specific weight of products of engineering areas and several other branches of industry produced with help of plastics grew from 32–35% in 1960 till 85–90% in 1990.

The specific weight of construction plastics among the total world production of plastics reached only 5% in 1975 [15]. But plastics of constructional use prevailed in world production of plastics in 1981–1985 [16].

The introduction of polymers has not only positive effect on the state of already existing traditional areas of industry but also determined the technical progress in rocket and atomic industries, aircraft industry, television, restorative surgery and medicine as a whole et cetera. The world production of construction polymers today is more than 10 million tons.

The thermal and heat-resistant constructional plastics take special place among the polymers. The need for such ones arises from fact that the use of traditional polymeric materials of technological assignment is limited by insufficient working thermal resistance, which is usually less than 103–150 °C.

Two classes of polymers have been used for producing high-tonnage thermally resistant plastics: aromatic polyarylates and polyamides [17].

Starting with these and mutually complementing the properties, materials of high operation properties, and thermo-stable plastics of constructional assignment in the first turn, have been produced.

Some widely used and attractive classes of polymers of constructional assignment are considered in detail below.

7.3 AROMATIC POLYESTERS OF *N*-OXYBENZOIC ACID

The *n*-hydroxybenzoic (*n*-oxybenzoic) acid has been extensively used at polymeric synthesis for the improvement of thermal stability of polymers for the last years [18, 19]. The aromatic polyester "ECONOL," homo-polymer of poly *n*-oxybenzoic acid, possesses the highest thermal resistant among the all homopolymeric polyethers [20] and attracts attention for the industry.

The poly *n*-oxybenzoic acid is the linear high-crystalline polymer with decomposition temperature in inert environment of 550 °C [20]. Below this point decomposition goes extremely slow. The loss in weight at 460 °C in air is 3% during 1 h of thermal treatment and than number is 1% at 400 °C. The CO, CO_2 and phenol are released when decomposing in vacuum at temperatures 500–565 °C, the coke remnant is of almost polyphenylene-like structure. It is assumed [27], that decomposition starts from breaking of ester bonds. The energy of activation of the process of fission of up to 30% of fly degradation products is 249,5 kJ/mol. The following mechanism for the destruction of polymer has been suggested:

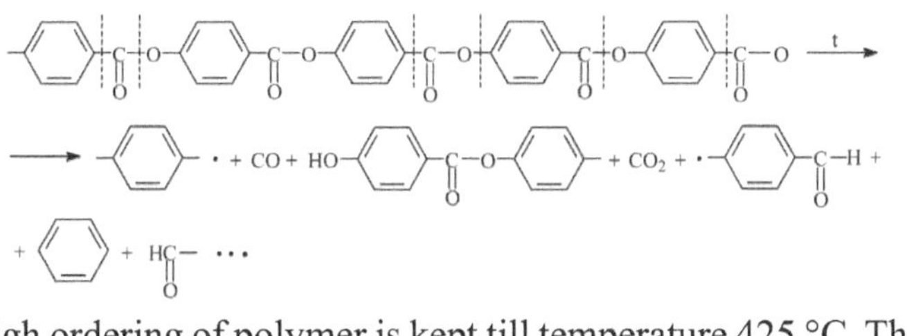

The high ordering of polymer is kept till temperature 425 °C. The equilibrium value of the heat of melting of polymers of *n*-oxybenzoic acid is found to be 5.4 kJ/mol [21].

Several methods for the producing the poly *n*-oxybenzoic acid have been proposed [20, 23–30], which are based of the high-temperature polycondensation, because the phenol hydroxyl possesses low reaction ability. Usually more reactive chloranhydrides are used. The polycondensation of *n*-oxybenzoic acid with blocked hydroxyl group (or of corresponding chloranhydride) happens at temperature of 150 °C and greater. The *n*-acetoxybenzoic acid [23, 24] or the *n*-oxybenzoic acid mixed with acylation agent Ac_2O [25–27] at presence of standard catalyst (in two stages: in solution at temperature 180–280 °C and in solid phase at temperature 300–400 °C) as well as chloranhydride of *n*-oxybenzoic acid [34], or the reaction of *n*-oxybenzoic acid with acidic halogenating agent (e.g., $SOCl_2$, PCl_3, PCl_5) [29] have been used at synthesis of high-molecular polymer aiming to increase the reactivity.

Besides the *n*-oxybenzoic acid, it's replaced in core derivatives (with F, Cl, Br, J, Me, Et etc. as replacers) can be used as initial materials in polycondensation processes. However, the low-molecular polymer of logarithmic viscosity equal to 0,16 is formed at polycondensation of methyl-replaced *n*-oxybenzoic aced in presence of triphenylphosphite [37] which is associated to the elapsing of adverse reaction of intramolecular etherification with triphenyl phosphite, resulting in termination of chain. This reaction does not flow when using PCl_3. High heat- and thermo-stable polyesters of *n*-oxybenzoic acid of given molecular weight containing no edge COOH-groups can be obtained by heating of *n*-oxybenzoic acid mixed with dialkylcarbonate at 230–400 °C [20].

The high-molecular polymer can be produced from phenyl ester of *n*-oxybenzoic acid at presence of butyltitanate in perchloroligophenylene at heating in current nitrogen during 4 h at temperatures 170–190 °C and later after 10 h of exposure to 340–360 °C [20]. Usually [32], Ti, Sn, Pb, Bi, Na, K, Zn or their oxides, salts of acetic, chlorohydratic or benzoic acid are used as catalysts of polycondensation processes when producing polyesters. The polycondensation of polymers of oxybenzoic acids and their derivatives has been performed at absence of catalysts [33] in current nitrogen at temperatures 180–250 °C and pressure during 5–6 h either. The temperature of polycondensation can be lowered [34] if it is carried out in polyphosphoric acid or using activating COOH– group of the substance.

Interesting investigation on study of structure characteristics of polyesters and copolyesters of *n*-oxybenzoic acid have been performed in Refs. [35, 36]. The introduction of links of *n*-oxybenzoic acid results in high-molecular compounds of ordered packing of chains, similar to the structure of high-temperature hexagonal modification of homopolymer of *n*-

oxybenzoic acid [37, 38]. The presence of fragments of *n*-oxybenzoic acid in macromolecular chain of the polymer not only increases the thermal resistance but improves the physic-chemical characteristics of polymeric materials. Obtained polyesters, consisting of monomers of *n*-oxybenzoic acid and *n*-dioxyarylene (of formula HO-Ar-OH, where Ar denotes bisphenylene, bisphenylenoxide, bisphenylensulfone), possess higher breaking impact strength, than articles from industrial polyester [39]. The data on thermogravimetric analysis of homopolymer of *n*-oxybenzoic acid and its polyesters with 4,4′-dioxybisphenylpropane, tere- or isophthalic acids are presented in Ref. [21]. The temperature of 10% loss of mass is 454, 482, 504 °C for them correspondingly.

Highly durable, chemically inert, thermo-stable aromatic polyesters can be produced by interaction of polymers containing links of *n*-oxybenzoic acid, aromatic dioxy-compounds, for example of hydroquinone and aromatic dicarboxylic acids [40]. Thermo and chemically resistant polyesters of improved mechanical strength can be obtained by the reaction of *n*-oxybenzoic acid, aromatic dicarboxylic acids, aromatic dioxy-compounds and diaryl carbonates held in solid phase or in high-boiling solvents [41] at temperature 180 °C and lowered pressure, possibly in the presence of catalysts [42]. Some characteristics of aromatic polyesters based on *n*-oxybenzoic acid are gathered in Table 7.1 [20].

As mentioned above, the thermal stability of polymers is closely tied to the manifestation of fire-protection properties. Many factors causing the stability of the materials to the exposure of high temperatures are characteristic for the fire-resistant polymers too.

Thermally stable polyesters can be produced on the basis of *n*-oxybenzoic acid involving stabilizer (triphenylphosphate) introduced on the last stage. The speed of the weight loss decreases two times, after 3 h of exposure to 500 °C polymer loses 0.87% of mass [43].

The high strength high-modular fiber with increased thermo- and fire-resistance can be formed from liquid-crystal copolyester containing 5–95 mol % of *n*-oxybenzoic acid.

The phosphorus is used as fire-resistant addition, in such a case the oxygen index of the fiber reaches 65% [44].

A 40–70 mol % of the compound of formula Ac-*n*-C_6H_4COOH are used to produce complex polyesters of high mechanical strength [45]. The polyesters with improved physic-chemical characteristics can be obtained by single-stage polycondensation of the melt of 30–60 mol % of *n*-oxybenzoic acid mixed with other ingredients [46].

TABLE 7.1 Physic-Mechanical Properties of Aromatic Polyesters Based on *N*-Oxybenzoic Acid

Composition of polyesters	Heat resistance on vetch, C, °C	Bending strength, MPa	Flex modulus, N/sm^2	Breaking strength, MPa	Elasticity modulus when break, N/sm^2
n-oxybenzoic acid, terephthalic acid	-	-	-	35.5	35,000
4,4′-bisphenylquinone				157	47,000
n-oxybenzoic acid, isophthalic acid, hydroquinone, 4,4′-benzophenon dicarboxylic acid	141	163	10,270	-	-
n-oxybenzoic acid, isophthalic acid, hydroquinone, bisphenylcarbonate	-	-	493	120.5	-
n-oxybenzoic acid, isophthalic acid, hydroquinone, 4,4′-bis-hydroxydiphenoxide or bisphenol S	130	160	6100	-	-
n-oxybenzoic acid, isophthalic acid, hydroquinone, 3-chlor-*n*-oxybenzoic acid and bisphenol D	133	232	10,260	-	-

The most used at synthesizing aromatic polyesters are the halogen-replaced anhydrides of dicarboxylic acids and aromatic dioxy-compounds of various constitutions. However, the growing content of halogens in polyesters obtained from mono-, bis- or tetra-replaced terephthalic acid [46] results in lowering of temperatures of glassing, melting as well as of the degree of crystallinity and to some decrease in mechanical strength for halogen-replaced bisphenols. Chlorine- and bromine-containing antipyrenes worsen the thermal resistance of polyesters and are usually used in a company with stabilizers [47, 48]. (Antipyrenes are chemical substances, either inorganic or organic, containing phosphorus, halogen, nitrogen, boron, and metals that are used for the lowering of the flammability of polymeric materials).

Besides, the usage of halogen-replaced bisphenols and dicarboxylic acids for synthesizing polyesters of lowered flammability considerably increases the cost of the latter. Consequently, the aromatic inhibiting flame additions are necessary for creating thermally and fire resistant polymeric materials. The introduction of such agents in small amounts would not diminish properties of polyesters.

Accounting for the aforesaid, one can assume that the chemical modification of known thermally resistant polyesters, by means of introduction of solider component (halogen-containing *n*-oxybenzoic acid) can help to solve the problem of increasing of fire safety of polyesters without worsening of their properties. It is obvious [49] that to replace one should intentionally use bromine atoms, which are more effective in conditions of open fire compared to atoms of other halogens.

So, as it comes from the above, the reactive compound involving halogens are widely used for imparting fire-protection properties to aromatic polyesters, as well as oligomer and polymeric antipyrenes.

The liquid-crystal polyesters (LQPs) became quite popular within the last decades. Those differ in their ability to self-arm, possess low coefficient of linear thermal expansion, have extreme size stability, are very chemically stable and almost do not burn.

LQPs can be obtained by means of polycondensation of aromatic oxy-acids (*n*-oxybenzoic one), dicarboxylic acids (iso- and terephthalic ones) and bisphenols (static copolymers) and also by peretherification of polymers and monomers.

Depending on the degree of ordering, LQPs are classified as smectic, nematic and cholesterol ones. Compounds, able to form liquid-crystal state, consist of long flat and quite rigid, in respect to the major axis, molecules.

LQPs can be obtained by several cases:

- Polycondensation of dicarboxylic acids with acetylic derivatives of aromatic oxy-acids (*n*-oxybenzoic one) and bisphenols.
- Polycondensation of phenyl esters of aromatic oxy-acids (*n*-oxybenzoic one) and aromatic dicarboxylic acids with bisphenols.
- Polycondensation of dichloranhydrides of aromatic dicarboxylic acids and bisphenols.
- Copolycondensation of dicarboxylic acids, diacetated of bisphenols and/or acetated of aromatic oxy-acids (*n*-oxybenzoic one) with polyethyleneterephthalate.

Liquid crystal copolyesters have been synthesized [50–56] on the basis of *n*-acetoxybenzoic acid, acetoxybisphenol, terephthalic acid and m-acetoxybenzoic acid by means of polycondensation in melt. All the copolyesters are thermotropic and form nematic phase. The types of LQPs of *n*-oxybenzoic acid are given in Table 7.2.

LQPs can be characterized by high physic-chemical parameters, see Table 7.3.

The fiber possessing high strength properties is formed from the melt of such copolyesters at high temperatures.

The synthesis of totally aromatic thermo-reactive complex copolyesters based on *n*-oxybenzoic acid, terephthalic acid, aromatic diols and alyphatic acids has been performed by means of polycondensation in melt [57–61].

Complex copolyesters are nematic static copolyesters.

The analysis of patents shows that various methods have been proposed for the production of liquid-crystal complex copolyesters [62–64].

TABLE 7.2 Label Assortment of Thermotropic Liquid-Crystal Polyesters of *N*-Oxybenzoic Acids

Company	Country	Trade label	Remarks
Dartco Manufacturing Inc.	USA	Xydar	Polyester based on *n*-oxybenzoic acid, terephthalic acid and *n*, *n*-bisphenol
		SRT – 300	Unfilled with normal fluidity
		SRT – 500	Unfilled with high fluidity
		FSR – 315	50% of talca
		MD – 25	50% of glass fiber
		FC – 110	Glass-filled
		FC – 120	Glass-filled
		FC – 130	Mineral filler
		RC – 210	Mineral filler
		RC – 220	Glass-filled
LNP Corp.		Thermocomp	Xydar + 150 polytetrafluorethylene (antifriction)
RTP Co		FDX – 65194	Xydar compositions: glass-filed with finishing and thermal-stabilizing additions, mineral fillers
Celanese Corp.		Vectra	Polyesters based on *n*-oxybenzoic acid and naphthalene derivatives

TABLE 7.2 *(Continued)*

Company	Country	Trade label	Remarks
Celanese Corp.		A – 130	30% of glass fiber
		B – 130	30% of glass fiber
		A – 230	30% of carbon fiber
		B – 230	30% of glass fiber
		A – 540	40% of mineral filler
		A – 900	Unfilled
		A – 950	Unfilled
Eastman Kodak Co	USA	Vectron	Copolyesters based on *n*-oxybenzoic acid and polyethyleneterephthalate
		LCC – 10108	
		LCC – 10109	
BASF	Germany	Ultrax	Re-replaced complex aromatic polyester
Bayer A.G.	Germany	Ultrax	Aromatic polyester
ICI	Great Britain	Victrex	Polyester based on *n*-oxybenzoic acid and acetoxynaphtoic acid
		SRP – 1500G	Unfilled
		SRP – 1500G–30	27% of glass fiber
		SRP – 2300G	Unfilled (special design)
		SRP – 2300G-30	27% of glass fiber (special design)
Sumimoto kagaku koge K.K	Japan	Ekonol – RE – 6000	Polyester based on *n*-oxybenzoic acid, isophthalic acid and bisphenol
Japan Elano Co		Ekonol	Fiber
Mitsubishi chemical Co		Ekonol	Aromatic polyester
Unitica K.K.		LC – 2000	Polyesters based on *n*-oxybenzoic acid and terephthalic acid
		LC – 3000	
		LC – 6000	

TABLE 7.3 Physic-Mechanical Properties of Some Liquid-Crystal Polyesters of *N*-Oxybenzoic Acid

Property	Xydar			Vectra				
	SRT-300	SRT-350	FSR-315	A-625 (chem. stable)	A-515 (highly fluid)	A-420 (wear resistant)	A-130 (filled with glass fiber)	C-130 (highly thermal resistant)
Density, g/cm^3	1.35	1.35	1.4	1.54	1.48	1.88	1.57	1.57
Tensile limit, MPa	115.8	125.5	81.4	170	180	140	200	165
Modulus in tension, GPa	9.65	8.27	8.96	10	12	20	17	16
Izod impact, J/m								
of samples with cut	128.0	208	75	130	370	100	135	120
of samples without cut	390.0	186	272	—	—	—	—	—
Tensile elongation, %	4.9	4.8	3.3	6.9	4.4	1.3	2.2	1.9
Deformation heat resistance (at 1,8 MPa), °C	—	—	—	185	188	225	230	240
Heat resistance on vetch, °C	366	358	353	—	—	—	—	—
Arc resistance, sec	138	138	—	—	—	—	—	—
Dielectric constant at 10^6 Hz	3.94	3.94	—	—	—	—	—	—
Dielectric loss tangent at 10^6 Hz	0.039	0.039	—	—	—	—	—	—

The description of methods for production of copolymers (having repetitive links from derivatives of *n*-oxybenzoic acid, 6-oxy-2-naphthoic acid, terephthalic acid and aromatic diol) formable from the melt is reported in Refs. [65–68]. Each repetitive link is in certain amounts within the polymer. All these copolymers are used as protective covers.

The thermotropic liquid-crystal copolyesters can be synthesized also on the basis of *n*-oxybenzoic and 2,6-oxynaphthoic acid at presence of catalysts (sodium and calcium acetates) [69, 70]. The catalyst sodium acetate accelerates the process of synthesis. Calcium acetate accelerates the process only at high concentration of the catalyst and influences on the morphology of complex copolyesters.

The syntheses of copolymers of *n*-oxybenzoic acid with polyethyleneterephthalate and other components are possible too [71–75]. It is established that copolymers have two-phase nature: if polyethyleneterephthalate is introduced into the reaction mix then copolymers of block-structure are formed.

The properties of liquid-crystal copolyesters based on *n*-oxybenzoic acid and oxynaphthoic acid (in various ratios and temperatures) are described in Refs. [76–81]. It was shown therein that the plates from such polymers formed by die-casting are highly anisotropic which results in appearance of the layered structure. Those nematic liquid crystals had ferroelectric ordering. Following substances are characterized in Refs. [82–85]: aromatic copolyesters of *n*-oxybenzoate/bisphenol A; those based on *n*-oxybenzoic acid, bisphenol and terephthalic acid; based on 40 molar % *n*-oxybenzoic acid, 30 molar % *n*-hydroquinone and isophthalic acid; and those based on *n*-oxybenzoic acid, *n*-hydroquinone and 2,6-naphthalenedicarboxylic acid. The influence of temperature, warm-up time and heating rate on the properties of copolyesters was studied: the glass-transition temperature was shown to increase with warm-up expanded. The number of works is devoted to the study of the complex of physic-chemical properties of liquid-crystal copolyesters based on *n*-oxybenzoic acid and polyethyleneterephthalate [86–97]. It was established that mixes up to 75% of liquid-crystal compound melt and solidify like pure polyethyleneterephthalate. Copolyesters, containing less than 30% of *n*-oxybenzoic acid, are in isotropic glassy state while copolyesters of higher concentration of the second compound are in liquid-crystal phase and can be characterized by bigger electric inductivity, than in glassy state. This distinction in caused by the existence of differing orientational distribution of major axes in relation to the direction of electric field in various structure instances of copolyesters. The data of IR-spectroscopy reveal that components of the mix interact in melt by means of reetherification reaction. Increasing pressure, decreasing free volume and mobility in the mix, one can delay the reaction between the compounds.

The investigation of the properties of liquid-crystal copolyesters based on *n*-oxybenzoic acid/polyethyleneterephthalate and their mixes with isotactic polypropylene (PP), polymethylmethacrylate, polysulfone, polyethylene-2,6-naphthalate, copolyester of *n*-oxybenzoic acid 6,2-oxynaphthoic acid continues in Refs. [98–106]. Specific volume, thermal expansion coefficient α and compressibility β were measured for PP mixed with liquid-crystal copolymer *n*-oxybenzoic acid/polyethyleneterephthalate. The high pressure results in appearance of ordering in melted PP. The increase of α and β for all mixes studied (25–100% PP) has been observed at temperatures about melting point of PP having those parameters for liquid-crystal copolyester changed inconsiderably. Increasing content of liquid-crystal copolyester in the mix remarkably decreases the value of α both in solid state and in area of melting PP. The latter has to be accounted for when producing articles. The liquid-crystal compound is grouped in shape of concentric cylinders of various radii in mixes with polysulphones. It becomes responsible for the viscous properties of the mixes at above conditions.

Piezoelectric can be produced from liquid-crystal polymers based on *n*-oxybenzoic acid/polyethyleneterephthalate and *n*-oxybenzoic acid/oxynaphthoic acid. Time-stability and temperature range of efficiency of piezoelectric from polymer made of 6,2-oxynaphthoic acid are higher compared to that of polyethyleneterephthalate.

It was found that the use of 30% liquid-crystal copolyester of *n*-oxybenzoic acid/polyethyleneterephthalate makes polymethylmethacrylate 30% more durable and increases the elasticity modulus on 110%, with the reprocessibility being the same.

The structure and propertied of various liquid-crystal copolymers are studied in Refs. [107–112]. For example, copolymers based on *n*-oxybenzoic acid, polyethyleneterephthalate, hydroquinone and terephthalic acid, are studied in Ref. [107]: the introduction of the mix hydroquinone/terephthalic acid accelerates the process of crystallization and increases the degree of crystallinity of polyesters.

The existence of two structure areas of melts of copolyesters, namely of low- temperature one (where high-melting crystals are present in nematic phase) and high-temperature other (where the homogeneous nematic alloy is formed) is revealed when studying the curves of fluxes of homogeneous and heterogeneous melts of copolyesters based on polyethyleneterephthalate and acetoxybenzoic acid [108].

The curves of flux of liquid-crystal melt are typical for viscoplastic systems while the trend for the flow limit to exist becomes more evident with increasing molar mass and decreasing temperature. Essentially higher quantities of molecular orientation and strength correspond to extrudates obtained from the homogeneous melt compared to extrudates produced from heterophase melt. The influence of high-melting crystallites on the process of disorientation of the structure and on the worsening of durability of extrudates becomes stronger with increasing contribution of mesogenic fragment within the chain.

When studying structure of liquid-crystal copolymers based on *n*-oxybenzoic acid/polyethyleneterephthalate and *m*-acetoxybenzoic acid by means of IR-spectroscopy, ^{1}H nuclear magnetic resonance and large angle X-ray scattering, the degree of ordering in copolymers was shown [109] to increase if concentration of *n*-oxybenzoic units went from 60 till 75%.

When studying properties of copolymers *n*-oxybenzoate/ethyleneterephthalate/*m*-oxybenzoate with help of thermogravimetry [110] the thermo-decomposition of copolymers was found to happen at temperatures 450–457°C in N_2 and 441–447 °C in air. The influence of the ratio of *n*- and *m*-isomers was regarded, the coal yield at $T > 500$ °C found to be 42.6% and increasing with growing number of *n*-oxybenzoic units.

The mixing of melts poly 4,4′-oxybenzoic acid/polyethyleneterephthalate was studied in Ref. [111]: the kinetic characteristics of the mix became incompatible at reetherification.

The measurement of glassing temperature T_{glass} [112] of thermotropic liquid-crystal polyesters synthesized from 4-acetoxybenzoic acid (component *A*), polyethyleneterephthalate (component *B*) and 4-acetoxyhydrofluoric acid revealed that the esters could be characterized by two phases, to which two glassing temperatures correspond: T_{glass}=66–83°C and T_{glass}=136–140 °C. The lower point belongs to the phase enriched in *B*, while higher one – to phase enriched in *A*.

Liquid-crystal polymers are most applicable among novel types of plastics nowadays. The chemical industry is considered to be the one of perspective areas of using liquid-crystal polymers where they can be used, because of their high thermal and corrosion stability, for replacement of stainless steel and ceramics.

Electronics and electrotechnique are considered as promising areas of application of liquid-crystal polymers. In electronics, however, liquid-crystal polymers meet acute competition from cheaper epoxide resins and polysulfone. Another possible areas of using high-heat-resistant liquid crystal polymers are avia, space and military technique, fiber optics (cover of optical cable and so on), auto industry and film production.

The question of improving fire-resistance of aromatic polyesters is paid more attention last time. Polymeric materials can be classified on criterion of combustibility: noncombustible, hard-to-burn and combustible. Aromatic polyesters enter the combustible group of polymers self-attenuating when taken out of fire.

The considerable fire-resistance of polymeric materials and also the conservation of their form and sizes are required when polymers are exploited to hard conditions such as presence of open fire, oxygen environment, exposure to high-temperature heat fluxes.

On the assumption of placed request, the extensive studies on both syntheses of aromatic polyesters of improved fire-resistance and modification of existing samples of polymers of given type have been carried out. The most used methods of combustibility lowering are following:

- coating by fireproof covers;
- introduction of filler;
- directed synthesis of polymers;
- introduction of antipyrenes;
- chemical modification.

The chemical modification is the widely used, easily manageable method of improving the fire resistance of polymers. It can be done synthetically, simultaneously with the copolymerization with reactive modifier via, for example, introduction of replaced bisphenols, various acids, other oxy-compounds. Or it can be done by addition of reactive agents during the process of mechanic-chemical treatment or at the stage of reprocessing of polymer melt [113].

The most acceptable ways of modification of aromatic polyesters aimed to get self-attenuating materials of improved resistance to aggressive media are the condensation of polymers from halogen-containing monomers, the combination of aromatic polyesters with halogen-containing compounds and the use of halogen-involving coupling agents [114, 115].

Very different components (aromatic and aliphatic) can become modifiers. They inhibit the processes of combustion and are able to not only to attach some new features to polymers but also improve their physical properties.

7.4 AROMATIC POLYESTERS OF TEREPHTHALOYL-BIS-(*N*-OXYBENZOIC) ACID

Among known classes of polymers the aromatic polyesters with rigid groups of terephthaloyl-bis-(*n*-oxybenzoic) acid in main chain of next formula attract considerable attention:

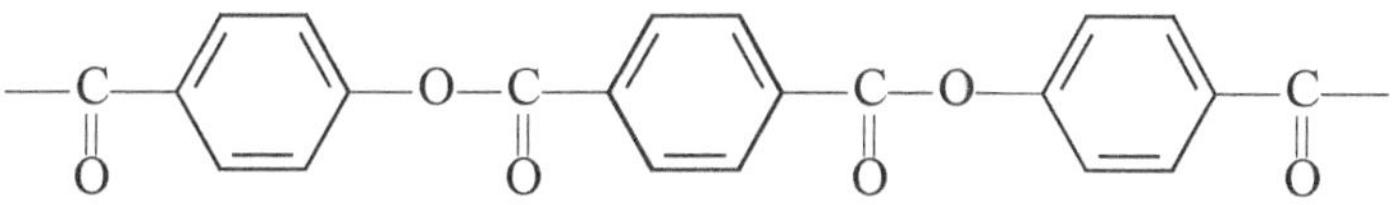

The polymers with alternating terephthaloyl-bis(*n*-oxybenzoatomic) rigid (R) mesogenic groups and flexible (F) decouplings of various chemical structure within the main chain (RF-copolymers) are able to form liquid-crystal order in the melt. The interest to such polymers is reasoned by fact that dilution of mesogenic “backbone” of macromolecule by flexible decouplings allows one to change the temperature border of polymer transfer from the partially crystalline state into the liquid-crystal one and also to change the interval of existence of liquid-crystal melt. The typical representatives of such class of polymers are polyesters with methylene flexible decouplings.

These polymers have been synthesized by high-temperature polycondensation of terephthaloyl-bis(*n*-oxybenzoylchloride) with appropriate diols in high-boiling dissolvent under the pressure of inert gas. Bisphenyloxide is used as dissolvent.

The features of conformation, orientation order, molecular dynamics of mentioned class of polymers are studied in Refs. [116–119] on the basis of polydecamethyleneterephthaloyl-bis(*n*-oxybenzoate), P-10-MTOB, which range of liquid-crystallinity is from 230 °C till 290 °C.

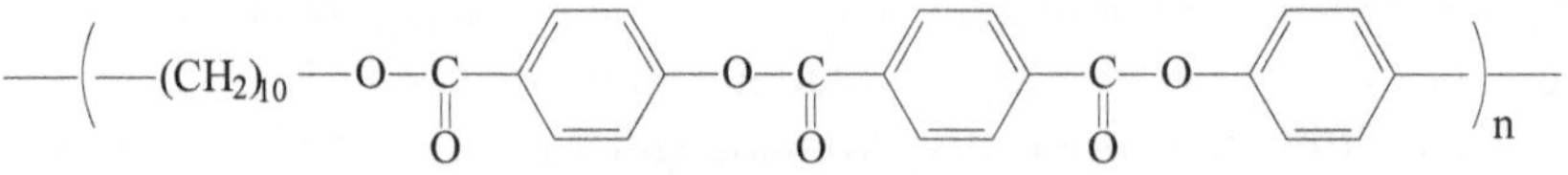

The study of polyesters with terephthaloyl-bis(*n*-oxybenzoate) groups continues in Refs. [120–123]. The oxyethylene $(CH_2CH_2O)_n$ and oxypropylene $(CH_2CHCH_3O)_n$ groups are used as flexible decouplings. It was found for oxyethylene decouplings that mesophase did not form if flexible segment was three times longer than rigid one. The folding of flexible decoupling was found to be responsible for the formation of ordered mesophase in polymers with long flexible decoupling. It was observed that polymers are able to form smectic and nematic liquid-crystal phases, the diapason of existence of which is determined by the length of flexible fragments.

The reported in Ref. [124] were the data on polymers containing methylene siloxane decouplings in their main chain:

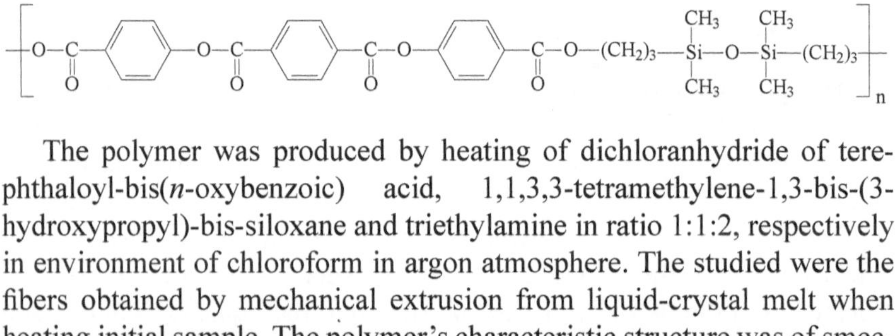

The polymer was produced by heating of dichloranhydride of terephthaloyl-bis(*n*-oxybenzoic) acid, 1,1,3,3-tetramethylene-1,3-bis-(3-hydroxypropyl)-bis-siloxane and triethylamine in ratio 1:1:2, respectively in environment of chloroform in argon atmosphere. The studied were the fibers obtained by mechanical extrusion from liquid-crystal melt when heating initial sample. The polymer's characteristic structure was of smectic type with folding location of molecules within layers.

The study of liquid-crystal state was performed in Ref. [125] on polymers with extended (up to 5 phenylene rings) mesogenic group of various lengths of flexible oxyethylene decouplings:

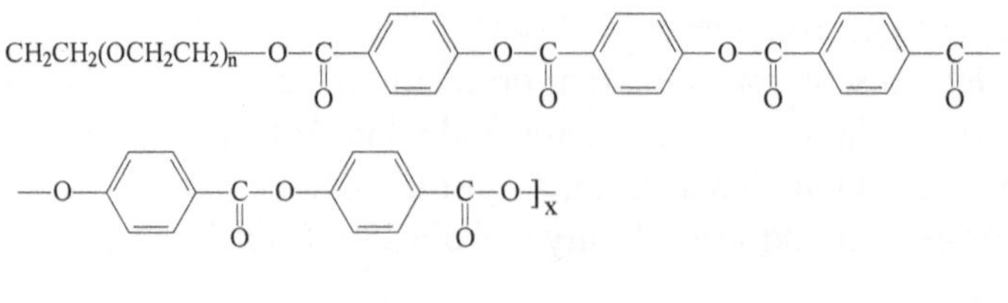

All polymers were obtained by means of high-temperature acceptor-free polycondensation of terephthaloyl-bis(*n*-oxybenzoylchloride) with bis-4-oxybenzoyl derivatives of corresponding polyethylene glycols in solutions of high-boiling dissolvent in current inert gas. Copolyesters with mesogenic groups, extended up to 5 phenylene cycles and up to 15–17 oxyethylene links, are able to form the structure of nematic type.

Polymers can also form the systems of smectic type [126]. The studied in Ref. [127] was the mesomorphic structure of polymer with extended group polyethylene glycol-1000-terephthaloyl-bis-4-oxybenzoyl-bis-4′-oxybenzoyl-bis-4″-oxybenzoate:

This polymer (of formula $R{=}CH_2CH_2(OCH_2CH_2)_{18-20}$) was synthesized by means of high-temperature acceptor-free polycondensation of terephthaloyl-bis-4-oxybenzoate with bis-(4-oxybenzoyl-4′-oxybenzoyl) derivative polyethylene glycol-1000 in the environment of bisphenyloxide [128]. The X-ray diffraction pattern revealed that the regularity of layered order weakened with rising temperature while ordering between mesogenic groups in edge direction kept the same. That type of specific liquid-crystal state in named polymer formed because of melting of layers including flexible oxyethylene decouplings at keeping of the ordering in transversal direction between mesogenic groups.

The Ref. [129] was devoted to the study of the molecular mobility of polymer, which formed mesophase of smectic type in temperature range 223–298 °C. Selectively deuterated polymers were synthesized to study the dynamics of different fragments of polymer in focus. That polymer was found to have several coexisting types of motion of mesogenic fragment and decoupling at the same temperature. There were massive vibration of phenylene cycles with varying amplitude and theoretically predicted movements of polymethylene chains [130]: trans-gosh-isomerization and translation motions involving many bonds. Such coexistence was determined by the phase microheterogeneity of polymers investigated.

The process of polycondensation of terephthaloyl-bis(*n*-oxybenzoylchloride) with decamethyleneglycol resulting in liquid-crystal polyester was studied in Ref. [131]. The monomers containing groups of complex esters entered into the reaction with diols at sufficiently high temperature (190 °C).

The relatively low contribution of adverse reactions compared to the main one led to regular structure of final polyester. The totally aromatic polyesters of following chemical constitution were studied in Ref. [132]: polymer "ФТТ-40" consisting of three monomers (terephthaloyl-bis(*n*-oxybenzoic) acid, terephthalic acid, phenylhydroquinone and resorcinol) entering to the polymeric chain in quantities 2:3:5. Terephthalic acid was replaced with monomer containing phenylene cycle in meta-place in polymer "ФГР-80."

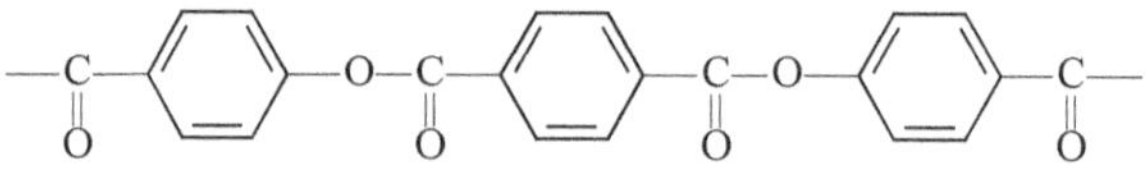

terephthaloyl-bis(*n*-oxybenzoic) acid (*A*),

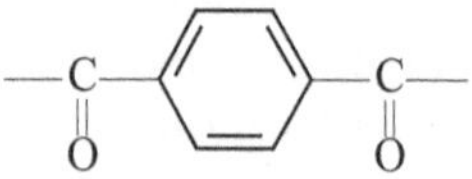

terephthalic acid (*B*),

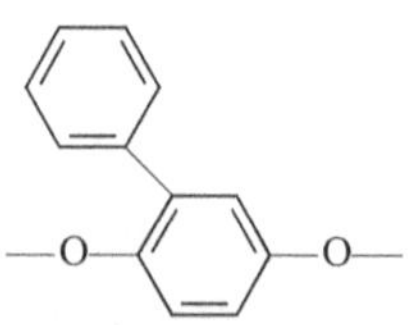

phenylhydroquinone (*C*),

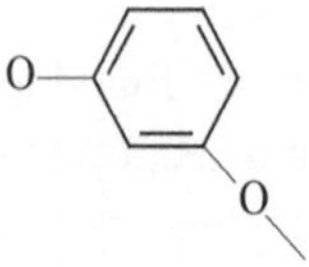

resorcinol (*D*).

The monomers entered to the polymeric chain in ratio *A*:*C*:*D*=5:4:1. These polymers melted at temperatures 300–310°C and switched to liquid-crystal state of nematic type. The monomers in "ФТТ-40" were found to be distributed statistically.

The influence of chemical structure of flexible decouplings in polyterephthaloyl-bis(*n*-oxybenzoates) on their mesogenic properties was studied in Ref. [131]. The studied were the polymers which had asymmetrical centers introduced into the flexible methylene decoupling and also complex ester groups:

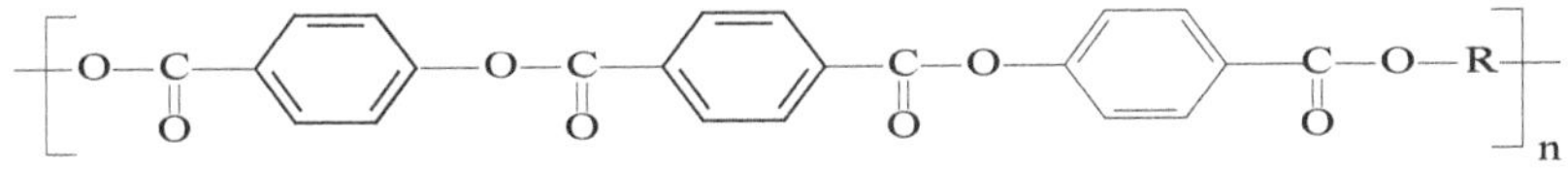

R=$CH(CH_3)CH_2CH_2CH_2CH_2$ (I); R=CH(R')$COOCH_2CH_2$(II, I)

where R'=$CH(CH_3)_2$ (II) or $CH_2CH_2(CH_3)$ (III).

The polymer I was synthesized by means of high-temperature acceptor-free polycondensation from dichloranhydride of terephthaloyl-bis(*n*-oxybenzoic) acid and 2-methylhexa-methylene-1,5-diol in inert dissolvent (bisphenyloxide at 200 °C). The phase state of polymer I at room temperature was found to depend on the way of sample preparation. The samples obtained from polymer immediately after synthesis and those cooled from melt were in mesomorphic state, but dried from the solutions in trifluoroacetic acid were in partially crystalline state.

The polymers II and III were in partially crystalline state at room temperature irrespective from the method of production.

So, the introduction of asymmetrical center into the flexible pentamethylene decoupling did not change the type of the phase state in the melt and did not influence on the temperature of the transitions into the area of existing of liquid-crystal phase. However, increasing of the rigidity of methylene decouplings via the introduction of complex-ester groups and enlarging of the volume of side branches considerably influenced on the character of intermolecular interaction what resulted in the formation of mesomorphic state of 3D structure during the melting of polymers.

The conformational and optic properties of aromatic copolyesters with links of terephthaloyl-bis(*n*-oxybenzoic) acid, phenylhydroquinone and resorcinol, containing 5% (from total number of para-aromatic cycles) of m-phenylene cycles within the main chain were studied in Ref. [134]. The polymer had the structure:

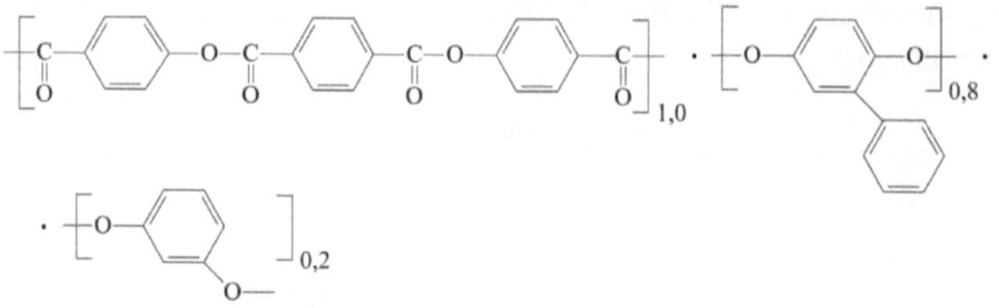

It was determined that the length of statistic Kuhn segment A was 200±20Å, the degree of dormancy of intramolecular rotations $(\sigma^{-2})^{1/2}=1.08$, the interval of molar masses started with 3.1 and ended $29.9\ \ 10^3$.

The analysis of literature data shows that the use of bisphenyl derivatives as elements of structure of polymeric chain allows one to produce liquid crystal polyesters, reprocessible from the melt into the articles of high deformation-strength properties and high heat resistance.

The polyethers containing mesogenic group and various bisphenylene fragments were synthesized in Ref. [135]:

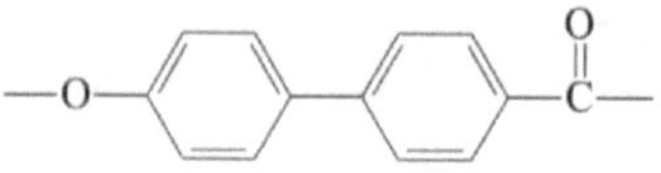

links of 4-oxybisphenyl-4-carbonic acid,

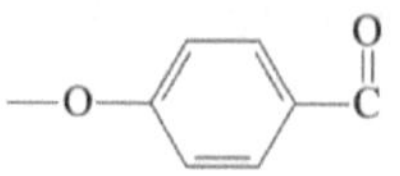

links of *n*-oxybenzoic acid,

links of terephthalic acid,

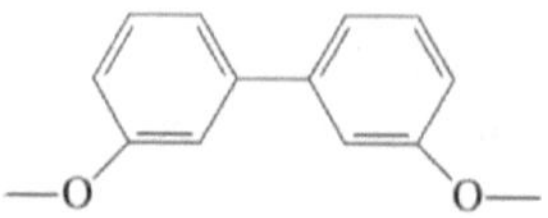

links of dihydroxylic bisphenyls of (3,3′-dioxybisphenyl).

All were linked with various decouplings within the main chain of next kinds:

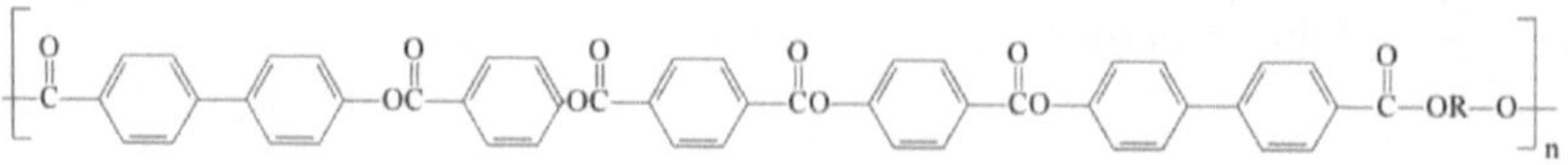

I

poly(alkyleneterephthaloyl-bis-4-oxybisphenyl-4′-carboxylate)

$[\eta] = 0.78$ dL/g $T_{glass} = 160$ °C, $T_{vapor} = 272$ °C, where R = $(CH_2)_6$; $(CH_2)_{10}$

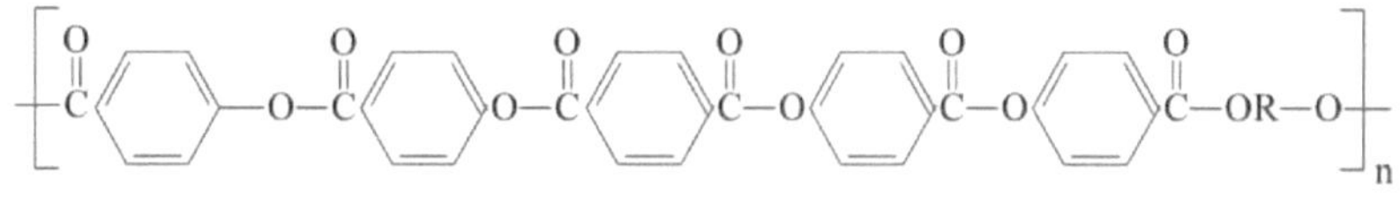

II

poly(oxyalkyleneterephthaloyl-bis-4-oxybenzoyl-4′-oxybenzoate, where R = $-CH_2-CH_2OCH_2-CH_2-$; $-(-CH_2CH_2O-)_2-CH_2CH_2-$ oxyethylene decouplings; polyethylene glycols (PEG) PEG 200, PEG 300, PEG 400, PEG 600, PEG 1000 – polyoxyethylene decouplings of various molecular weights and copolymers based on 3,3′-dioxybisphenyl with elements of regularity within the main chain:

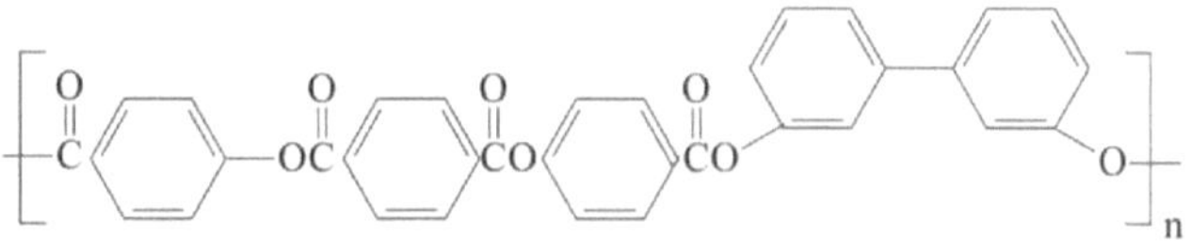

III

links of terephthaloyl-bis(*n*-oxybenzoic) acid – links of 3,3′-dioxybisphenyl

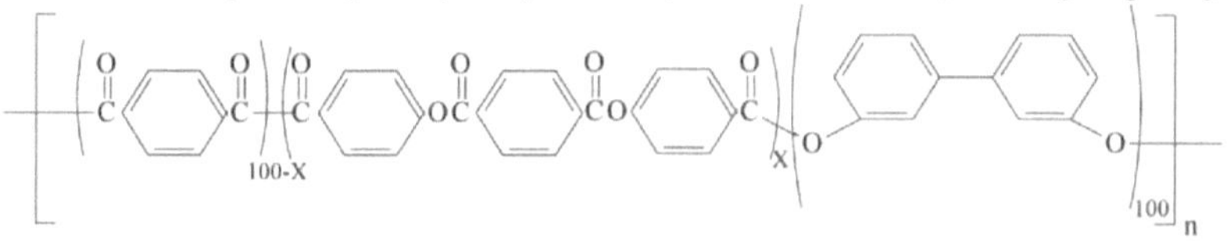

IV

links of terephthalic acid – links of terephthaloyl-bis(*n*-oxybenzoic) acid – links of 3,3′-dioxybisphenyl.

The synthesized in Ref. [136] were the block-copolymers containing the links of terephthaloyl-bis(*n*-oxybenzoate) and polyarylate. It was found that such polymers had liquid-crystal nematic phase and the biphase separation occurred at temperatures above 280 °C.

The thermotropic liquid-crystal copolyester of polyethyleneterephthalate and terephthaloyl-bis(*n*-oxybenzoate) was described in Ref. [137]. The snapshots of polarized microscopy revealed that copolyester was nematic liquid-crystal one.

When studying the reaction of reetherification happening in the mix 50:50 of polybutyleneterephthalate and complex polyester, containing mesogenic sections from the remnants of *n*-oxybenzoic (I) and terephthalic (II) acids separated by the tetramethylene decouplings the content of four triads with central link (I) and three triads with central link

II was determined in Ref. [138]. The distribution of triad sequences approximated to the characteristic one for statistical copolymer with increasing of the exposure time.

The relaxation of liquid-crystal thermotropic polymethyleneterephthaloyl-bis(*n*-oxybenzoate) was studied in Ref. [139]: the β-relaxation occurred at low temperatures while α-relaxation took place at temperatures above 20°C (β-relaxation is associated with local motion of mesogenic groups while α-relaxation is caused by the glassing in amorphous state). It was established thin the presence of ordering both in crystals and in nematic mesophase expanded relaxation.

So, the liquid-crystal nematic ordering is observed, predominantly, in considered above polymeric systems containing links of terephthaloyl-bis(*n*-oxybenzoic) acid [140] and combining features of both polymers and liquid crystals. Such ordering can be characterized by the fact that long axes of mesogenic groups are adjusted along some axis and the far translational ordering in distribution of molecules and links is totally absent.

The analysis of references demonstrates that polyesters with links of terephthaloyl-bis(*n*-oxybenzoic) acid can be characterized by the set of high physic-mechanical and chemical properties.

That is why we have synthesized novel copolymers on the basis of dichloranhydride of terephthaloyl-bis(*n*-oxybenzoic) acid and various aromatic oligoesters.

7.5 AROMATIC POLYARYLENESTERKETONES

The acceleration of technical progress, the broadening of assortment of chemical production, the increment of the productivity of labor and quality of items from plastics are to a great extent linked to synthesis, mastering and application of new types of polymeric materials. New polymers of improved resistance to air, heat form-stability, longevity appeared to meet the needs of air and space technique. Filling and reinforcement can help to increase the heat resistance on about 100 °C. Plastics enduring the burden in temperature range 150–180 °C are called "perspective" while special plastics efficient at 200 °C are named "exotic."

The most promising heat-resistant plastics for hard challenges are polysulfones. Today, the widely used aromatic polysulfones as constructional and electroisolating materials are those of general formula $[-O-Ar-SO_2-]_n$ where R stands for aromatic radical.

The modified aromatic polysulfones attract greater interest every day. The introduction of ester group into the chain makes the molecule flexible, elastic, more "fluid" and reprocessible. The introduction of bisphenylsulphonic group brings thermal resistance and form stability. Non-modified polysulfones are hardly reprocessible due to the high temperature of melt and are not used for technical purposes.

Aromatic polysulfones are basically linear amorphous polyerylensulfonoxides. They are constructional thermoplasts, which have sulfonic groups $-SO_2-$ in their main chain along with simple ester bonds, aliphatic and aromatic fragments in different combinations.

There are two general methods known for the production of polyarylenesulfonoxides from hydroxyl-containing compounds [8, 141]. These are the polycondensation of disodium or dipotassium salts of bisphenols with dahalogenbisphenylsulfone and the homopolycondensation of sodium or potassium salts of halogenphenoles. One should notice that mentioned reactions occur according to the bimolecular mechanism of nucleophylic replacement of halogen atom within the aromatic core.

The first notion of aromatic polysulfones as promising constructional material was met in 1965 [142–145]. The most pragmatic interest among aromatic polysulfones attracts the product of polycondensation of 2,2-bis(-4-oxyphenyl) of propane and 4,4′-dichlorbisphenyldisulfone:

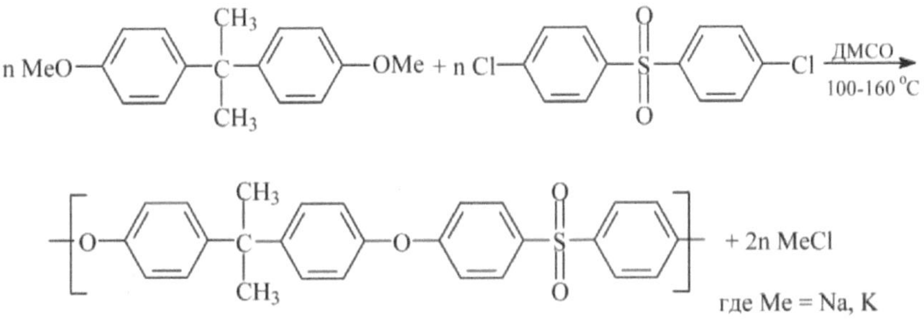

The degree of polymerization of industrial polysulfones varies from 60 till 120 what corresponds to the weights from 30,000 to 60,000.

The other bisphenols than bisphenylolpropane (diane) can be used for synthesizing polysulfones, and the constitution of ingredients make considerable effect on the properties of polymers.

Linear polysulfones based on bisphenylolpropane and containing isopropylidene groups in the chain are easily reprocessible into the articles and have high hydrolysis stability. The presence of simple ester links in the polymeric chains makes them more flexible and durable. The main effect on properties of such polysulfones is produced by sulfonic bond which makes the polymer more stable to oxidation and more resistant to heat. Above properties of polysulfones along with the low cost of bisphenylolpropane change them into almost ideal polymers for constructional plastics. Polysulfones of higher heat-resistance can be obtained on the basis of some other bisphenols.

The industrial production of polysulfones based on 2,2-bis-(4oxyphenyl) of propane and 4,4′-dichlorbisphenylsulfone was started by company "Union Carbid" (USA) in 1965 [146].

The manufacture of given aromatic polysulfones was also launched by company "ICI" in Great Britain under the trademark "Udel" in 1966. Various labels of constructional materials (P-1700, P-1700-06, P-1700-13, P-1700-15, etc.) are developed on the basis of this type of polysulfones.

Aromatic polysulfones are soluble in different solvents: good dissolution in chlorized organic solvents and partial dilution in aromatic hydrocarbons.

By its thermo-mechanical properties the aromatic polysulfone on the basis of bisphenylolpropane takes intermediate place between polycarbonate and polyarylate of the same bisphenol. The glassing temperature of the polysulfone lies in the ranges of 190–195 °C, heat resistance on vetch is 185 °C. The given polysulfone is devised to be used at temperatures below 150 °C and is frost-resistant material (–100 °C).

One of the most valuable properties of polysulfones is well creep resistance, especially at high temperatures. The polysulfone's creep deformation is 1.5% at 100 °C and after 3.6×10^6 sec loading of 21 MPa, which is better than that of others. Their long-term strength at high temperature is also better. Thus, the polysulfone can be used as constructional material instead of metals.

The flow limit of polysulfone is 71.5 MPa, the permanent strain after rupture is 50–100%. At the same time the elasticity modulus (25.2 MPa) points on sufficient rigidity of material which is comparable to that of polycarbonate. Impact strength on Izod is 7–8 kJ/m^2 at 23 C with notch.

The strength and durability of polysulfones keep well at high temperatures. This fact opens possibilities to compete with metals in such areas where other thermoplasts are worthless.

The possibilities of using polysulfones in parts of high-precision articles arise from its low shrinkage (0.7%) and low water absorption (0.22% after 5.64×10^4 sec).

Aromatic polysulfone on the basis of bisphenylolpropane is relatively stable to thermo-oxidation destruction, because the sulfur is in its highest valence state in such polymers; electrons of adjacent benzene nuclei shift, under the presence of sulfur, to the side of sulfogroups what causes the resistance to oxidation.

The polysulfone can operate long at temperature up to 140–170 °C, the loss of weight after 2.52×10^7 sec of loading at 125 °C is less than 0.25%; polymer loses 3% of its mass after 3.24×10^7 sec of exposure to 140 °C in oxygen.

The results of thermal destruction of polysulfone in vacuum have revealed that the first product of decomposition at 400°C was the sulfur dioxide; there are also methane and bisphenylpropane in products of decomposition at temperatures below 500 °C. Therefore, the polysulfone is one of the most stable to thermo-oxidation thermoplasts.

The articles from polysulfone possess self-attenuating properties, caused by the nature of polymer but not of the additives. The values of oxygen index for polysulfone lie in the range of 34–38%. Apparently, aromatic polysulfones damp down owing to the formation of carbonized layer, becoming porous protective cover, on their surfaces. The probable is the explanation according to which the inert gas is released from the polymer [8].

Aromatic polysulfone is chemically stable. It is resistant to the effect of mineral acids, alkalis and salt solutions. It is even more stable to carbohydrate oils at higher temperatures and small loadings [149].

The polysulfone on the basis of bisphenylpropane also possesses high dielectric characteristics: volume resistivity 10^{17}; electric inductivity at a frequency 10^6 Hz is 3, 1; dielectric loss tangent at a frequency 10^6 Hz is 6×10^{-4}.

Aromatic polysulfones can be reprocesses by die-casting, by means of extrusion, pressing, blow method [150, 151]. The aromatic polysulfones should be dried out for approximately 1.8 10^4 sec at 120 °C until the moisture is more than 0.05%. The quality of articles based on polysulfone worsens at higher moists though the polymer itself does not change its properties. The polysulfones are reprocessible at temperatures 315–370 °C.

The high thermal stability of aromatic polysulfones allows one to conduct multitime reprocessing without the destruction of polymer and loss of properties. The pressed procurements can be mechanically treated on common machine tools.

Besides the aromatic polysulfone on the basis of bisphenylpropane, the other aromatic polysulfones are produced industrially. In particular, the industrial manufacture of polysulfones has been carried out by company "Plastik 3 M" (USA) since 1967 (the polymers are produced by reaction of electrophylic replacements at presence of catalyst). The polysulfone labeled "Astrel-360" contains links of following constitution:

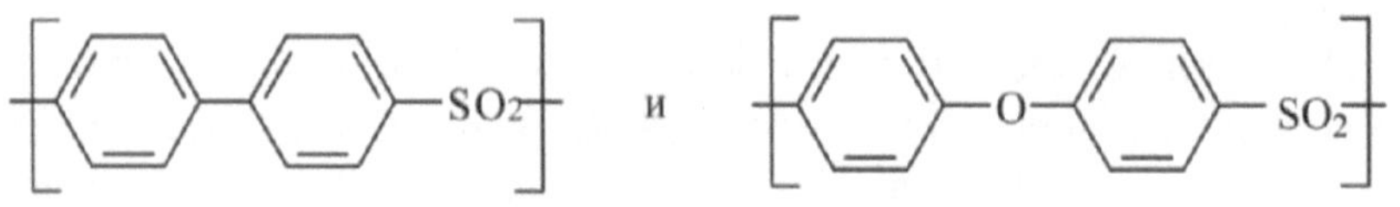

The larger amount of first-type links provides for the high glassing temperature (285 °C) of the polymer. The presence of second-type links allows one to reprocess this thermoplast with help of pressing, extrusion and die casting on special equipment [152].

The company "ICI" (Great Britain) produced polysulfones identical to "Astrel-360" in its chemical structure. However, the polymer "720 P" of that firm contains greater number of second-type links. Due to this, the glassing temperature of the polymer is 250 °C and its reprocessing can be done on standard equipment [153].

Pure aromatic structure of given polymers brings them high thermo-oxidation stability and deformation resistance. Mechanical properties of polysulfone allow one to classify them as technical thermoplasts having high durability, rigidity and impact stability. The polymer "Astrel 360" is in the same row with carbon steel, polycarbonate and nylon on its durability.

The polysulfone has following main characteristics: water absorption – 1.8% (within 5.64 × 10^4 sec); tensile strength – 90 MPa (20 °C) and 30 MPa) (260 °C); bending strength – 120 MPa (20 °C) and 63 MPa (260 °C); modulus in tension – 2.8 GPa (20 °C); molding shrinkage – 0.8%; electric inductivity at a frequency 60 Hz – 3.94; dielectric loss tangent at 60 Hz – 0.003 [154–157].

The polysulphones have good resistance to the effect of acids, alkalis, engine oils, oil products and aliphatic hydrocarbons.

The polysulfone "Astrel-600" can be reprocesses at harder conditions. Depending on the size and shape of the article, the temperature of pressed material should lie within the ranger of 315–410 °C while pressure should be about 350 MPa [158]. The given polysulfone can be reprocessed by any present method. Articles from it can be mechanically treated and welded. Thanks to its properties, polysulfone finds broad use in electronique, electrotechnique and avia-industry [159, 160].

The Refs. [161–163] reported in 1968 on the production of heat-resistant polyarylenesulfonoxides "Arilon" of following constitution (company "Uniroyal," USA):

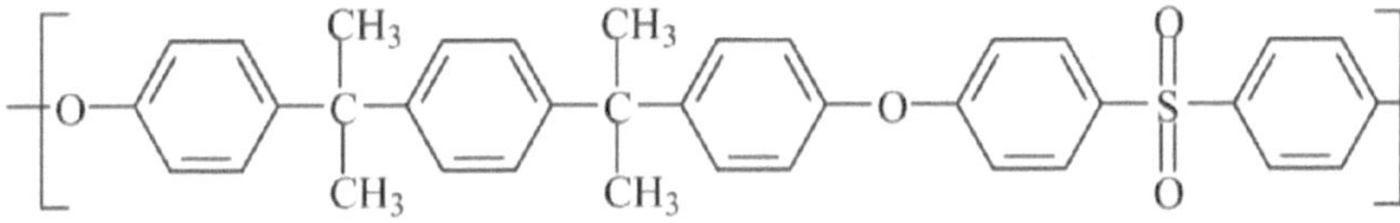

"Arilon" has high rigidity, impact-resistance and chemical stability. The gain weight was 0.9% after 7 days of tests in 20% HCl and 0.5% in 10% NaOH.

This polymer is suitable for long-term use at temperatures 0–130 C and can be easily reprocessed into articles by die casting. It also can be extruded and exposed to vacuum molding with deep drawing.

Given polysulfone finds application in different fields of technique as constructional material.

Since 1972, the company "ICI" (Great Britain) has been manufacturing the polyarylenesulfonoxide named "Victrex" which forms at homopolycondensation of halogenphenoles [164–166] of coming structure formula:

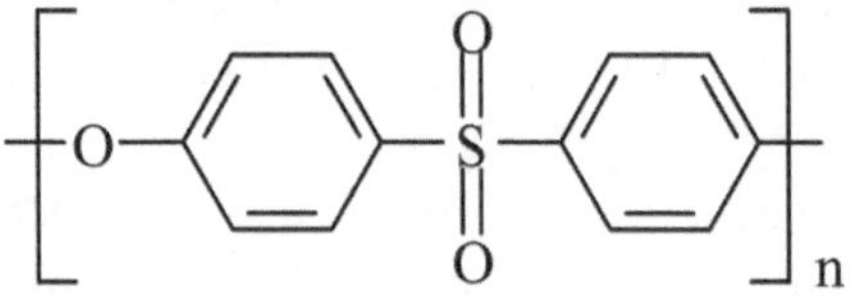

Several types of polysulfone "Victrex" can be distinguished: 100P – powder for solutions and glues; 200P – casting polymer; 300P – with increased molecular weight for extrusion and casting of articles operating under the load at increased temperature in aggressive media; and others.

The polysulfone "Victrex" represents amorphous thermoplastic constructional polymer differing by high heat-resistance, dimensional staunchness, low combustibility, chemical and radiation stability. It can easily be reprocessed on standard equipment at 340–380 °C and temperature of press-form 100–150 °C. It is dried at 150 °C for 1.08×10^4 sec before casting [167, 168].

This polysulfone is out of wide distribution yet. However, it will, presumably, supplant the part of nonferrous metal in automobile industry, in particular, when producing carburetors, oil-filters and others. The polysulfone successively competes with aluminum alloys in avia industry: the polymer is lighter and yields neither in solidity, neither in other characteristics.

The company "BASF" (Germany) has launched the production of polyethersulfone labeled "UltrasonE" [169] representing amorphous thermoplastic product of polycondensation; it can be characterized by improved chemical stability and fire-resistance. The pressed articles made of it differ in solidity and rigidity at temperature 200 °C. It is assumed to be expedient to use this material when producing articles intended for exposure to increased loadings when the sizes of the article must not alter at temperatures from −100 °C till 150 °C. These items are, for instance in electrotechnique, coils formers, printing and integrated circuits, midspan joints and films for condensers.

"UltrasonE" is used for producing bodies of pilot valves and shaped pieces for hair dryers.

The areas of application of polysulfones are extremely vary and include electrotechnique, car – and aircraft engineering, production of industrial, medical and office equipment, goods of household purpose and packing.

The consumption of polysulfones in Western Europe has reached 50% in electronics/electrotechnique, 23% in transport, 12% in medicine, 7% in space/aviation and 4% in other areas since 1980 till 2006 [170].

The polysulfones are used for manufacturing of printed-circuit substrates, moving parts of relays, coils, clamps, switches, pipes socles, potentiometers details, bodies of tools, alkaline storage and solar batteries, cable and capacitor insulation, sets of television and stereo-apparatuses, radomes. The details under bonnet, the head lights mirrors, the flasks of hydraulic lifting mechanisms of cars are produced from polysulfones. They are also met in internal facing of planes cockpits, protective helmets of pilots and cosmonauts, details of measuring instruments.

The polysulfones are biologically inert and resistant to steam sterilization and γ-radiation what avails people to use them in medicine when implanting artificial lens instead of removed due to a surgical intervention and when producing medical tools and devices (inhalers bodies, ophthalmoscopes, etc.).

The various methods of synthesizing of polysulfones have been devised by Russian and abroad scientists within the last 10–15 years.

The bisphenylolpropane (diane), 4,4′-dioxybisphenylsulfone, 4,4′-dioxybisphenyl, phenolphthalein, hydroquinone, 4,4′-dioxyphenylsulfonyl – bisphenyl are used as initial monomers for the synthesis of aromatic polysulfones. The polycondensation is carried out at temperatures 160–320 °C with dimethylsulfoxide, dimethylacetamide, N-methylpirrolydone, dimethylsulfone and bisphenylsulfone being used as solvents [171–174].

The Japanese researchers report [175–179] on the syntheses of aromatic polysulfones via the method of polycondensation in the environment of polar dissolvent (dimethylformamide, dimethylacetamide, and dimethylsulfoxide) at 60–400 °C in the presence of alkali metal carbonates within 10 min to 100 h. The synthesized thermoplastic polysulfones possess good melt fluidity [175].

The durable thermo-stable polysulfones of high melt fluidity can be obtained by means of polycondensation of the mix of phenols of 1,3-bis(4-hydroxy-1-isopropylidenephenyl and bisphenol A with 4,4′-dichlordimethylsulfone in the presence of anhydrous potassium carbonate in the environment of dimethylformamide at temperature 166 °C. The solution of the polymer is condensed in MeOH, washed by water and dried out at 150 °C in vacuum. The polysulfone has η_{limit} = 0.5 dL/g (1% solution in dimethylformamide at 25 °C) [176].

The aromatic polysulfones with the degree of crystallinity of 36% can be synthesized [179] with help of reaction of 2,2-bis(4-hydroxy-4-tret-butylphenyl) propane with 4,4′-dichlorbisphenylsulfone in the presence of potassium carbonate in the environment of polar solvent 1,3-dimethyl-2-imidasolidinone in current nitrogen at temperatures 130–200 °C. After separation and purification polymer has $\eta_{limit} = 0.5$ dL/g.

The possibility to use the synthesized aromatic polysulfone as antipyrenes of textile materials was shown in Ref. [180].

The production of polysulfones was shown to be possible in Ref. [181] by means of interaction of sulfuric acid, sulfur trioxide or their mixes with aromatic compounds (naphthalene, methylnaphthalene, methoxynaphthalene, dibenzyl ester, bisphenylcarbonate, bisphenyl, stilbene) if one used the anhydride of carbonic acid at 30–200 °C as the process activator. Obtained polymers could be reprocesses by means of pressing,

The synthesis of polyarylenesulfones containing links of 1,3,5-triphenylbenzene can be performed [182] by oxidation of polyarylene thioesters. Obtained polyarylenesulfones have T_{glass} = 265–329 °C and temperature of 5% loss of weight of 478–535 °C. They are well soluble in organic solvents and possess fluorescent properties.

To produce the polysulfone with alyphatic main chain, the interaction of SO_2-group of vinyl-aromatic compound (e.g., of sterol) with nonsaturated compound (for instance of acrylonitrile) or cyclic olefin (such as 1,5-cyclooctadiene) has been provided [183] in the presence of initiator of the radical polymerization in bulk or melt, the reaction being carried out at temperatures from 80 to 150 °C.

The method of acceptor-catalyst polyetherification can be used to produce [184, 185] thermo-reactive polyarylenesulfones on the basis of nonsaturated 4,4′-dioxy-3,3′-diallylbisphenyl-2,2′-propane and various chlor- and sulfo-containing monomers and oligomers. The set of physic-mechanical properties of polymers obtained allows one to propose them as constructional polymers, sealing coatings, film materials capable of operating under the influence of aggressive environments and high temperatures.

Relatively few papers are devoted to the synthesis and study of polycondensational block-copolyesters. At the same time this area represents, undoubtedly, scientific and practical interest. The number of researches [186–190] deals with the problem of synthesis and study of physic-chemical properties of polysulfones of block constitution.

For example, it is reported in Ref. [186] on the synthesis of block-copolyesters and their properties in dependence on the composition and structure of oligoesters. The oligoesters used were oligoformals on the basis of diane with the degree of condensation 10 and oligosulfones on the basis of phenolphthalein with the degree of condensation 10. The synthesis has been performed in conditions of acceptor-catalyst polycondensation.

The polysulfone possesses properties of constructional material: high solidity, high thermo-oxidation stability.

However, the deficiency of polysulfonic material is the high viscosity of its melt what results in huge energy expenses at processing. One can decrease the viscosity if, for example, "sews" together polysulfonic blocks with liquid-crystal nematic structures which usually have lower values of viscosity. To produce the block-copolymer, the flexible block of polysulfone and rigid polyester block (with liquid-crystal properties) were used in Ref. [187]. The polysulfonic block was used as already ready oligomer of known molecular weight with edge functional groups.

Polyester block represented the product of polycondensation of phenylhydroquinone and terephthaloylchloride.

The synthesis of block-copolymer went in two stages. At first, the polyester block of required molecular weight was produced while on the second stage that block (without separation) reacted to the polysulfonic block. The synthesis was carried out by the method of high-temperature polycondensation in the solution at 250 °C in environment of α-chlornaphthalene. The duration of the first stage was 1.5 h, and the second stage was 1 h. Obtained polymers exhibited liquid-crystal properties.

The block-copolymers of polysulfone and polyesters can be also produced [188] by the reaction of aromatic polysulfones in environment of dipolar aprotone dissolvents (dimethylsulfoxide, N-methylpirrolydone, N-methylcaprolactam, N, N{}-dimethylacetamides or their mixes) with alyphatic polyesters containing not less than two edge OH-groups in the presence of basic catalyst – carbonates of alkali metals: Li, Na, K.

Polysulfone is thermo-mechanical and chemically durable thermoplast. But in solution, which is catalyzed by alkali, it becomes sensitive to nucleophylic replacements. In polar aprotone dissolvents at temperatures above 150 °C in the presence of spirit solution of K_2CO_3 it decomposes into the bisphenol A and diarylsulfonic simple esters. The analogous hydrolysis in watery solution of K_2CO_3 goes until phenol products of decomposition. This reaction is of preparative interest for the synthesis of segmented block-copolyester simple polyester – polysulfone. The transetherification of polysulfone, being catalyzed by alkali, results, with exclusion of bisphenol A and introduction segments of simple ester, in formation of segmented block-copolymer [189].

The Refs. [190–199] are devoted to the problem of synthesis of statistical copolymers of polysulfones, production of mixes on the basis of polysulfones and study of their properties as well to the mechanism of copolymerization.

The graft copolymer products, poly(met)acrylates branched to polyestersulfones, can be produced next way [200]. Firstly, the polyestersulfone is being chlormethylenized by monochlordimethyl ester. The product is used as macrostarter for the graft radical polymerization of methylmethacrylate (I), methylacrylate (II) and butylacrylate (III) in dimethylformamide according to the mechanism of transferring of atoms under the influence of the catalytical system $FeCl_2$/isophthalic acid. The branched copolymer with I has only one glassing temperature while copolymer with II and III has three.

Lately, several researches on the synthesis of liquid-crystal polysulfones and on the production of mixes and melts of polysulfones with liquid-crystal polymers have appeared [201–205].

For example, the polysulfone "Udel" can be modified [201] by means of introduction of chlormethyl groups. Then the reaction of transquarternization of chlormethylized polysulfone and obtained nitromethylene dimesogens is conducted. The dimesogens contain one phenol OH-group and form nematic mesophase in liquid state. Obtained such a way liquid-crystal polysulfones, having lateral rigid dimesogenic links, possess the structure of enantiotropic nematic mesophase fragments.

The Ref. [202] reports on the synthesis of liquid-crystal polysulfone with mesogenic link – cholesterylpentoatesulfone.

The effect of compatibility, morphology, rheology, mechanical properties of mixes of polysulfones and liquid-crystal polymers are studied in Refs. [203–205]. There are several contributions on the methods of synthesizing of copolymers of polysulfones and polyesterketones and on the production of mixes [206–210]. The method for the synthesis of aromatic copolyestersulfoneketones proposed in Ref. [206] allows one to decrease the number of components used, to lower the demands to the concentration of moist in them and to increase the safety of the process. The method is in the interaction without aseotropoformer in environment of dimethylsulfone of bisphenols, dihaloydarylenesulfones and (or) dihaloydaryleneketones and alkali agents in the shape of crystallohydrated of alkali metal carbonated and bicarbonates. All components used are applicable without preliminary drying.

The novel polyarylenestersulfoneketone containing from the cyclohexane and phthalasynone fragments is produced via the reaction of nucleophylic replacement of 1-methyl-4,5-bis(4-chlorbenzoyl)cyclohexane, 4,4′-dichlorbisphenylsulfone and 4-(3,5-dimethyl-4-hydroxyphenyl)-2,3phthalasine-1-one. The polymer is described by means of IR-spectroscopy with Fourier-transformation, ^{1}H nuclear magnetic resonance, differential scanning calorimetry and diffraction of X-rays. It is shown that the polymer is amorphous and has high glassing temperature (200 °C), it is soluble in several dissolvents at room temperature. The by-products are the "sewn" and graft copolymers [207].

The block copolymers of low-molecular polyesterketoneketone and 4,4′{}-bisphenoxybisphenylsulfone (I) can be produced [208] by the reaction of polycondensation. The glassing temperatures increase and melting points lower of copolymers if the concentration of (I) increases. Block copolymers, containing 32.63–40.7% of (I) have glassing temperature and melting point 185–193 and 322–346 °C, respectively; the durability and modulus in tension are 86.6–84.2 MPa and 3.1–3.4 GPa, respectively; tensile elongation reaches 18.5–20.3%. Block-copolymers possess good thermal properties and reprocessibility in melt.

The analysis of literature data and patent investigation has revealed that the production of such copolymers as polyestersulfoneketones, liquid-crystal polyestersulfones as well as the preparation of the mixes and melts of polysulfones with liquid-crystal polyesters of certain new properties are of great importance.

With the account for the upper-mentioned, we have synthesized polyestersulfones on the basis of oligosulfones and terephthaloyl-bis(*n*-oxybenzoic) acid [174, 186, 434] and polyestersulfoneketones on the basis of aromatic oligosulfoneketones, mixes of oligosulfones with oligoketones of various constitution and degree of polycondensation and different acidic compounds [211, 212, 435].

7.6 AROMATIC POLYSULFONES

The stormy production development of polymers, containing aromatic cycles like, for example, the polyarylenesterketones has happened within the past decades. The polyarylenesterketones represent the family of polymers in which phenylene rings are connected by the oxygen bridges (simple ester) and carbonic groups (ketones). The polyarylenesterketones include polyesterketone, polyesteresterketone and others distinguishing in the sequence of elements and ratio E/K (of ester groups to ketone ones). This ratio influences on the glassing temperature and melting point: the higher content of ketones increases both temperatures and worsens reprocessibility [213].

The elementary units of polysteresterketones contain two simple ester and one ketone groups, while those of polyesterketone – only one ester and one ketone [214].

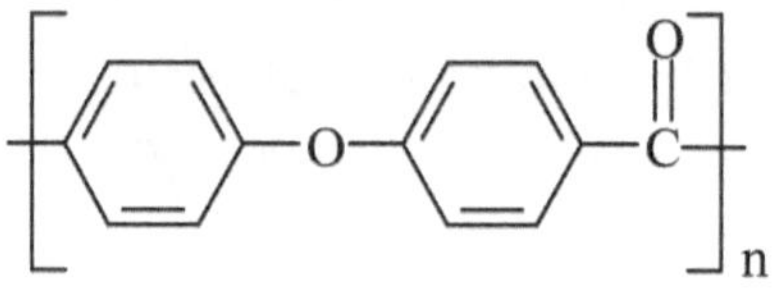

ПЭК

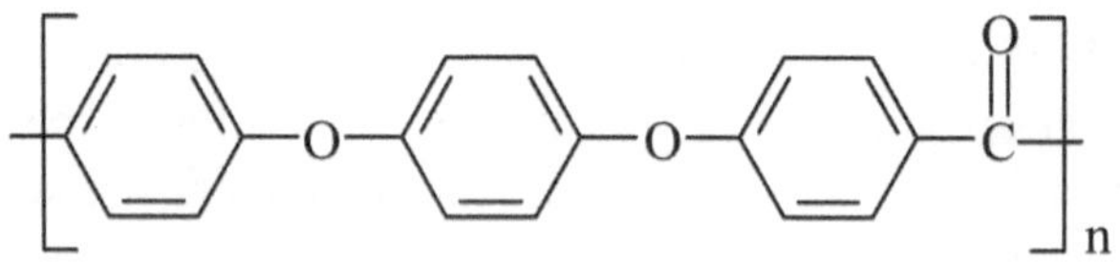

ПЭЭК

Polyesteresterketone is partially crystalline polymer the thermo-stability of which depends on glassing temperature (amorphosity) and melting point (crystallinity) and increases with immobilization of macromolecules. The strong valence bonds define the high thermo-stability and longevity of mechanical and electrical properties at elevated temperature.

The polyesterketone labeled "Victrex^{R}" was firstly synthesized in Great Britain by company "Imperial Chemical Industries" in 1977. The industrial manufacture of polyesteresterketones started in 1980 in Western Europe and USA and in 1982 in Japan [215].

The polyesteresterketone became the subject of extensive study from the moment of appearance in industry. The polyesteresterketone possesses the highest melting point among the other high-temperature thermoplasts (335 °C) and can be distinguished by its highly durable and flexible chemical structure. The latter consists of phenylene rings, consecutively joined by para-links to ester, ester and carboxylic groups.

There is a lot of information available now on the structure and properties of polyesteresterketones [216, 217].

The polyesteresterketone is specially designed material meeting the stringent requirements from the point of view of heat resistance, inflammability, products combustions and chemical resistance [218, 219]. "Victrex^{R}" owns the unique combination of properties: thermal characteristics and combustion parameters quite unusual for thermoplastic materials, high stability to effect of different dissolvents and other fluids [220, 221].

The polyesteresterketone can be of two types: simple (unarmored) and reinforced (armored) by glass. Usually both types are opaque though they can become transparent after treatment at certain conditions. This happens due to the reversible change of material's crystallinity which can be recovered by tempering. The limited number of tinges of polyesteresterketones has been produced for those areas of industry where color articles are used.

The structure crystallinity endows polyesteresterketones by such advantages as:

- stability to organic solvents;
- stability to dynamical fatigue;
- improved thermal stability when armoring with glass;
- ability to form plasticity at short-term thermal aging;
- orientation results in high strength fibers.

"Victrex^{R}" loses its properties as elasticity and solidity moduli with increasing temperature, but the range of working temperatures of polyesteresterketones is wider in short-term process (like purification) than that of other thermoplastic materials. It can be exploited at 300 □C or higher. The presumable stint at 250 °C is more than 50,000 hours for the given polyesteresterketone. If one compares the mechanical properties of polyesteresterketone, polyestersulfone, nylon and polypropylene then he finds that the first is the most resistant to wear and to dynamic fatigue. The change of mechanical characteristics was studied in dependence of sorption of CH_2CCl_2.

Polyesteresterketones, similar to any other thermoplastics, is isolation material. It is hard-burnt and forms few smoke and toxic odds in combustion, the demand in such materials arises with all big hardening requirements to accident prevention.

If one compares the smoke-formation in combustion of 2–3 mm of samples from ABC-plastics, polyvinylchloride, polystyrene, polycarbonate, polytetrafluorethylene, phenolformaldehyde resin, polyestersulfone, polyesteresterketone then it occurs that the least smoke is released by polyesteresterketone, while the greatest amount of smoke is produced by ABC-plastic.

"Victrex^{R}" exhibits good resistance to water reagents and pH-factor of different materials starting with 60% sulfuric acid and 50% potassium hydroxide. The polyesteresterketone dissolves only in proton substances (such as concentrated sulfuric acid) or at the temperature close to its melting point. Only α-chlornaphthalene (boiling point 260 °C) and benzophenone influence on "Victrex^{R}" among organic dissolvents.

The data on the solubility have revealed that two classes of polyesteresterketones coexist: "amorphous" and crystalline [222].

The division of these polymers into two mentioned classes is justified only by that the last class, independent of condensation method, crystallizes so fast in conditions of synthesis that the filtering of combustible solution is not possible. It may be concluded from results obtained that "amorphous" class of polyesteresterketones is characterized by bisphenols, which have hybridized sp^3-atom between phenyl groups.

From the point of view of short-term thermal stability the polyesteresterketones do not yield most steady materials – polyestersulfones – destruction of which is 1% at 430 °C. Yet and still their long-term stability to UV – light, oxygen and heat must be low due to ketone-group [222].

The influence of environment on polyesteresterketones is not understood in detail, but it has proven that polyesteresterketone fully keeps all its properties within 1 year. Polyesteresterketones exhibit very good stability to X-ray, β- and γ- radiations. Wire samples densely covered with polyesteresterketones bear the radiation 110 Mrad without essential destruction.

The destroying tensile stress of polyesteresterketone is almost nil at exposure to air during 100 hours at 270 °C. At the same time the flex modulus at glassing temperature of 113 °C falls off precipitously, however remains sufficiently high compared to that of other thermoplasts.

When placed into hot water (80 °C) for 800 hours the tensile stress and the permanent strain after rupture of polyesteresterketones decreases negligibly. The polyesteresterketone overcomes all the other thermoplasts on stability to steam action. The articles from polyesteresterketone can stand short exposure to steam at 300 °C.

On fire-resistance this polymer is related to hard-to-burn materials.

The chemical stability of polyesterester ketone "Victrex[R]" is about the same as of polytetrafluorethylene while its long-term strength and impact toughness are essentially higher than those or nylon A-10 [223].

The manufacture of the polyesteresterketones in Japan is organized by companies "Mitsui Toatsu Chem" under the labels "Talpa-2000", "ICI Japan," "Sumitoma Kogaku Koge." The Japanese polyesteresterketones have glassing temperature 143 C and melting point 334 C [224, 225].

The consumption of polyesteresterketones in Japan in 1984 was 20 tons, 1 kilogram cost 17000 Ian. The total consumption of polyestersulfones and polyesteresterketones in Japan in 1990 was 450–500 tons per year [226].

Today, 35% of polyesteresterketones produced in Japan (of general formula $[-OC_6H_4-O-C_6H_4-CO-C_6H_4-]_n$) are used in electronics and electrotechnique, 25% – in aviation and aerospace technologies, 10% – in car manufacture, 15% – in chemical industry as well as in the fields of everyday life, for example at producing the buckets for hot water, operating under pressure and temperature up to 300 °C. The Japanese industrial labels of polyesteresterketones have good physical-mechanical characteristics: the high impact toughness; the heat resistance (152 °C, and 286 °C with introduction of 20% of glass fiber); the chemical resistance (bear the influence of acids and alkalis, different chemicals and medicines), the tolerance for radiation action; the elasticity modulus 250–300 kg/mm^2; the rigidity; the lengthening of 100%; the negligible quantity of smoke produced. The polyesteresterketone can be reprocessed by die casting at 300–380 °C (1000–1400 kg/cm^2), extrusion, formation and others methods [224]. High physical-mechanical properties remain the dame for the long time and decrease on 50% only after 10 years.

The polymer is expensive. One of the ways for reducing the product price is compounding. The company "Kogaku Sumitoma" created compounds "Sumiploy K" on the basis of polyesteresterketones by their original technology. The series "Sumiploy K" includes the polymers with high strength and of improved wear resistance. The series "Sumiploy SK" is based on the polyesteresterketone alloyed with other polymers [224]. The series includes polymers, the articles of which can be easily taken out from the form, and products are of improved wear resistance, increased high strength, good antistatic properties, can be easily metalized.

The company "Hoechst" (Germany) produces unreinforced polyesteresterketones "Hoechst X915," armored with 30 weight% of glass fiber (X925) and carbon fiber (X935), which are characterized by the good physical-mechanical properties (Table 7.4). The unreinforced polymer has the density almost constant till glassing temperature (about 160 °C). The reinforcing with fibers permits further to increase the heat resistance of polyesterester ketone. At the moment the polyesteresterketones labeled "Hostatec" are being produced with 10, 20 and 30 weight % of glass and carbon fibers. Several labels of polyesteresterketones are under development which involve mineral fillers, are not reinforced and contain 30 weight % of glass and carbon fibers.

The constructional thermoplast "Hostatec" dominates polyoxymethylene, polyamide and complex polyesters on many parameters.

The polyesteresterketones can be easily processed by pressing, die cast and extrusion. They can be repeatedly crushed to powder for secondary utilization. They are mainly used as constructional materials but also can be used as electroinsulation covers operating at temperatures of 200 C and higher for a long time [225].

TABLE 7.4 Physic-Mechanical Properties of Polyesteresterketones of Company "Hoechst"

Property	Unreinforced	Containing 30 weight %	
		Glass fiber	Carbon fiber
Density, g/cm^3	1.3	1.55	1.45
Linear shrinkage, %	1.5	0.5	0.1
Tensile strength, N/mm^2	86	168	218
Breaking elongation, %	3.6	2.2	2.0
Modulus in tension, kN/mm^2	4	13.5	22.5
Impact toughness notched, J/m	51	71	60
Heat resistance, °C	160	Above 320	320

The polyesteresterketone found application in household goods (in this case its high heat resistance and impact toughness are being used), in lorries (joint washers, bearings, probes bodies, coils and other details, contacting fuel, lubricant and cooling fluid).

The big attention to polyesteresterketones is paid in aircraft and space industries. The requirements to fire-resistance of plastics used in crafts have become stricter within the last years. Unreinforced polyesteresterketones satisfy these demands having the fire-resistance category U-O on UL 94 at thickness 0.8 mm. In addition, this polymer releases few smoke and toxic substances in combustion (is used in a subway). The polyesteresterketone is used for coating of wires and cables, used in details of aerospace facility (the low inflammability, the excellent permeability and the wear stability), in military facility, ship building, on nuclear power plants (resists the radiation of about 1000 Mrad and temperature of water steam 185 °C), in oil wells (pillar stand to the action of water under pressure, at a temperature of 288 °C), in electrical engineering and electronics. Polyesteresterketone offer properties of thermoreactive resin, can be easily pressed, undergoes overtone, resists the influence of alkalis.

Since January 1990, the Federal aviation authority have adopted the developed at Ohio State University method, in which the heat radiation (HR) and the rate of heat release (RHR) are determined. The standard regulates the HR level at 65 kJ. Many aircraft materials do not face this demand: for example, the ternary copolymer of acrylonitrile, butadiene and styrene resin, the polycarbonate, phenol and epoxide resins. It is foreseen to substitute these materials by such polymers, which meet these requirements. The measurements in combustion chamber of Ohio State University indicated that polyesteresterketone fulfills this standard.

The heightened activity in creating and evaluation of composite properties on the basis of polyesteresterketones occurs recent years [228]. The thermoplastic composites have the number of advantages regarding the plasticity, maintainability and ability for secondary utilization compared to epoxy composites. These polymers are intended to be applied in alleviated support elements. In the area of cable insulation, polyesteresterketones can be reasonably used when thermo-stability combined with fire-resistance without using of halogen antipyrenes where is desired.

The "Hostatec" has low water absorption. Dielectric properties of films from polyesteresterketone "Hostatec" are high. This amorphous polymer has the electric inductivity 3, 6, loss factor 10^{-3} and the specific volume resistance 10^{17} Ohm×cm; these values remain still up to 60 °C.

Growth of demand for polyesteresterketones is very intense. In connection with growing demands on heat resistance and stability to various external factors the polyesteresterketones find broader distribution. The cost of one kilogram of such polymer is 5–20 times larger than the cost of usual constructional polymers (polycarbonates, polyamides, and polyformaldehydes). But, despite the high prices the polyesteresterketones and compositions on their basis, owing to the high level of consumer characteristics, find more and more applications in all industries. The growth of production volumes is observed every year.

It is known, that the paces of annual growth of polyesteresterketones consumption were about 25% before 1995, and its global consumption in 1995 was 4000 tons. The polyesteresterketone do not cause ecological problems and is amenable to secondary processing.

In connection with big perspectiveness of polyarylesterketones, the examining of the most popular methods for their production was of interest. The literature data analysis shows that the synthesis of aromatic polyesterketones can be done by the acylation on Friedel-Crafts reaction or by reaction of nucleophylic substitution of activated dihalogen-containing aromatic compounds and bisphenolates of alkali metals [224, 229].

In the majority cases, the polyesterketones and polyesteresterketones are produced by means of polycondensation interaction of bisphenols with 4,4′-dihalogen-substituted derivatives of benzophenone [230–263], generally it is 4,4′-difluoro- or dichlorbisphenylketone. The introduction of replacers into the benzene ring of initial monomer raises the solubility of polyesterketones and polyesteresterketones. So, polyesterketone on the basis of 3,3′,5,5′-tetramethylene and 4,4′-difluorobenzophenone is dissolved at 25 °C [211–214] in dimethylsulfoxide, the reduced viscosity of melt with concentration 0.5 gram/deciliter is 0.79 dL/g.

The polyesterketone based on 4,4′-dioxybenzophenone and dichlormethylenized benzene derivatives is soluble in chloroform and dichloroethane [264]. The logarithmic solution viscosity of polyesterketone obtained on the basis of 4,4′-dioxybisphenylsulfone and 4,4′-dichlorbenzophenone in tetrachloroethane of concentration 0.5 gram/deciliter is 0.486; the film materials from this polyester (of thickness 1 mm) are characterized by the high light transmission (86%) and keep perfect solubility and initial viscosity after exposure to 320 °C during 2 h.

The high-boiling polar organic solvents – dimethylsulfoxide, sulfolane, dimethylsulfone, dimethylformamide, dimethylacetamide – are generally used for synthesizing the polyesterketones and polyesteresterketones by means of polycondensation; in this case the reaction catalysts are the anhydrous hydroxides, carbonates, fluorides and hydrides of alkali metals.

The polymers synthesis is recommended to be carried out in inert gas atmosphere at temperatures 50–450 °C. If catalysts used are the salts of carbonic or hydrofluoric acids then oligomers appear. Chain length regulators when producing polyesterketones based on difluoro- or dichlorbenzophenone and bisphenolates of alkali metals and dihydroxynaphthalenes can be the monatomic phenols [251–253].

The synthesis of polyesterketones and polyesteresterketones according to the Friedel-Crafts reaction is lead in mild conditions [265–283]. So, solidifying thermo-stable aromatic polesterketonesulfones, applied as binding agents when laminating, can be produced [278] in the presence of aluminum chloride by the interaction of 1,4-di(4-benzoylchloride)butadiene-1,3-dichloranhydrides of iso- and terephthalic acids, bisphenyl oxide and 4,4′-bisphenoxybisphenylsulfone.

The aromatic polyesterketones and their thioanalogs are synthesized [266–281, 284] with help of polycondensation of substituted and not substituted aromatic esters and thioesters with choric anhydrides of dicarboxylic acids in environment of aprotone dissolvents at temperatures from −10 till 100 °C in the presence of Lewis acids and bases.

The aromatic polyesterketones and polyesterketonesulfonamides based on 4,4′-dichloranhydride of bisphenyloxidebicarbonic acid and 4-phenoxybenzoylchloride can be produced by means of Friedel-Crafts polycondensation in the presence of $AlCl_3$ [270]. The reduced viscosity of the solution in sulfuric acid of concentration 0.5 gram/deciliter is 0.07–1.98 dL/g. The Friedel-Crafts reaction can also be applied to synthesize the copolyesterketones from bisphenyl ester and aromatic dicarboxylic acids or their halogenanhydrides [274]. The molecular mass of polymers, assessed on the parameter of melt fluidity, peaks when bisphenyl ester is used in abundant amount (2–8%).

The polyarylesterketones can be produced by means of interaction between bisphenylsulfide, dibenzofurane and bisphenyloxide with monomers of electrophylic nature (phosgene, terephthaloylchloride) or using homopolycondensation of 4-phenoxybenzoylchloride and 4-phenoxy-4-chlorcarbonyl-bisphenyl in the presence of dichloroethane at 25 °C [282–285]. Aromatic polyesterketones form after the polycondensation of 4-phenoxybenzoylchloride with chloranhydrides of tere- and isophthalic acids, 4,4′-dicarboxybisphenyloxide in the environment of nitrobenzene, methylchloride and dichloroethane at temperatures from −70 till 40 °C during 16–26 h according Friedel-Crafts reaction.

The synthesis of polyesterketones based on aromatic ether acids is possible in the environment of trifluoromethanesulfonic acid [249, 287]. The data of ^{13}C nuclear magnetic resonance have revealed [249] that such polyesterketone comprise only the *n*-substituted benzene rings. When using the N-cyclohexyl-2-pyrrolidone as a solvent when synthesizing polyphenylenesterketones and polyphenylenethioesterketones the speed of polycondensation and the molecular mass of polymers [288] increase.

So, the polymer is produced, the reduced viscosity of 0.5% solution of which in sulfuric acid is 1.0 dL/g, after the interaction of 4,4′-difluorobenzophenone and hydroquinone at 290 °C in the presence of potassium carbonate during 1 h. Almost the same results are obtained when synthesizing polyphenylenethioesterketones. However, the application of the mix bisphenylsulfone/sulfolane as a solvent during the interaction of 4,4′-difluorobenzophenone with sodium sulfide within 2–13 h provides for the production of polymer with reduced viscosity 0.23–0.25 dL/g. The high-molecular polyarylenesulfideketones, suitable for the preparation of films, fibers and composite materials are formed during the interaction of 4,4′-dihalogenbenzophenone with sodium hydrosulfide in solution (N-methylpirrolydone) at 175–350 C within 1–72 h [289].

For the purpose of improving physical-mechanical properties and increasing the reprocessibility of polyesterketones and polyesteresterketones their sulfonation by liquid oxide of hexavalent sulfur in environment of dichloroethane has been carried out [290]. In this case the polymer destruction does not happen, which is observed at sulfonation by concentrated sulfuric or chlorosulfonic acid. The abiding flexible film materials can be produced from sulfonated materials using the method of casting. The aromatic polyesterketones could also be produced by oxidative dihydropolycondensation (according to the Scolla reaction) of 4,4′-bis-(1-naphthoxy)-benzophenone at 20 °C in the presence of trivalent iron chloride in environment of nitrobenzene [291]. The mix of pentavalent phosphorus oxide and methylphosphonic acid in ratio 1:10 can be used as a solvent and dehydrate agent [292, 293] when synthesizing the aromatic polyesterketones on the basis of bisphenylester of hydroquinone, 4,4′-bisphenyloxybicarbonic acid, 1,4-bis(m-carboxyphenoxy) benzene, and also for homopolycondensation of 3- or 4-phenoxybenzoic acid at 80–140 °C.

The sulfur-containing analogs of polyesterketones, that is, polythioesterketones and copolythioesterketones can be synthesized [294–303] by polycondensation of dihalogenbenzophenols with hydrothiophenol or other bifunctional sulfur-involving compounds, and also of their mixes with different bisphenols in environment of polar organic solvents. As in case of polyesterketones, the synthesis of their thioanalogs is recommended to be carried to out in inert medium at temperatures below 400 °C in the presence of catalyst (hydroxides, carbonates and hydrocarbonates of alkali metals).

The aromatic polyesterketones can also be produced by polycondensation or homopolycondensation of compounds like halogen-containing arylketone phenols and arylenedihalogenides of different functionality at elevated temperature and when the salts of alkali and alkali-earth metals in environments of high-boiling polar organic solvents are used as catalysts [222, 304–312]. The 4-halogen-3-phenyl-4-hydroxybenzophenone, 4-(*n*-haloidbenzoyl)-2,6-dinethylphenol and others belong to monomers with mixed functional groups.

The polycondensation process when synthesizing polyesterketones and polyesteresterketones can be performed in the melt [223, 313–317] too. So, it is possible to produce the aromatic polyesterketones by means of interaction in melt of 4,4′-difluorobenzophenone with trimethylsiloxane esters of bisphenols with different bridged groups in the presence of catalyst (cesium fluoride – 0.1% from total weight of both monomers) at 220–270 °C [318]. The monomers do not enter the reaction without catalyst at temperatures below 350 °C. The reduced viscosity of 2% solution of polymer in tetrachloroethane at 30 °C is 0.13–1.13 dL/g, the molecular weight – 3200–60,000, glassing temperature is 151–186 °C, melting point is 240–420 °C. According to the data of thermogravimetric analysis in the air the mass loss of polymer is less than 10%, when temperature elevates from 422 °C to 544 °C with the rate 8°/min.

To increase the basic physical-mechanical characteristics and reprocessibility (the solubility, in particular), the polyesterketones and polyesteresterketones are synthesized [319–329] through the stage of formation of oligomers with end functional groups accompanied with the consequent production of block-copolyesterketones or by means of one-go-copolycondensation of initial monomers with production of copolyesterketones.

The polyesteresterketones, their copolymers and mixes are used for casting of thermally loaded parts of moving transport, instrument, machines, and planes. They are used in articles of space equipment: for cable insulation, facings (pouring) elements. For instance, the illuminators frames of planes and rings for high-frequency cable are made of polyesterketone "Ultrapek." It is widely used in electronics, electrician, for extrusion of tubes and pipes operating in aggressive media and at low temperatures. The polyesteresterketones and polyester ketones are used for multilayer coating as the basis of printed planes. The conservation of mechanical strength in conditions of high humidity and temperature, the stability to radiation forwards their application to aerospace engineering.

The compositions on the basis of polyesteresterketones already compete with those based on thermoreactive resins when making parts of military and civilian aircrafts. Sometimes it is possible to cut the weight on 30% if produce separate parts of plane engines from reinforced polyesteresterketones.

It is proposed to use the polyesteresterketones in the manufacture of fingers of control rod and cams of brake system, motors buttons in car industry. The piston cap of automobile engine, made from polyesteresterketone "Victrex," went through 1300 h of on-the-road tests.

The advantage of using this material, contrary to steel, lays in wearout decrease, noise reduction, 40% weight reduction of the article. The important area of application of polyesteresterketones can be the production of bearings and backings. The polyester ester ketones are recommended to be used for piece making of drilling equipment (zero and supporting mantles) and timber technique, in different joining's of electric equipment of nuclear reactors, layers and valves coverings, components of sports facility.

The fibers of diameter 0.4 mm (from which the fabric is weaved in shape of tapes and belts used in industrial processes where the temperature-resistant, the high-speed conveyers are needed) are produced from polyester ester ketone melt at temperatures of 350–390 °C. The fabric from polyesteresterketone or polyesterketone keeps 90% of the tensile strength after thermal treatment at 260 °C, does not change its properties after steaming at 126 °C for 72 h under the load, and resists alkalis action with marginal change. The medical instruments, analytical, dialysis devices, endoscopes, surgical and dental tools, containers made of polyarylesterketones can be sterilized by steam and irradiation [330].

The manifold methods of synthesizing polyarylenesterketones have been devised by Russian and foreign scientists within the last 20 years.

For example, the method for production of aromatic polyketones by means of Friedel-Craft polycondensation of bis(arylsilanes) with chlorides of aromatic dicarboxylic acids (isophthaloyl-, terephthaloyl-, 4,4{}-oxydibenzoylchloride) at 20 °C in environment of dissolvent (1,2-dichlorethane) in the presence of aluminum chloride is proposed in Ref. [331]. The polyketones have the intrinsic viscosity more than 0.37 dL/g (at 30 °C, in concentrated H_2SO_4), glass transition temperature is 120–231 °C and melting point lies within 246–367 °C. The polyketones start decomposing at a temperature of 400 °C, the temperature of 10% loss of mass is 480–530 °C.

It is possible to produce polyketones by the reaction of aromatic dicarboxylic acid and aromatic compound containing two reactive groups [332]. The reaction is catalyzed by the mix of phosphoric acid and carboxylic acid anhydride having the formula of RC(O)O(O)CR (R stands for not-substituted or substituted alkyl, in which one, several or all hydrogen atoms were replaced by functional groups and each R has the Gamet constant $\sigma_m \geq 0.2$). The pressed articles can be created from synthesized polyketones.

The aromatic polyketones can be synthesized [333] when conducting the Friedel-Craft polymerization with acylation by means of the interaction of 2,2{}bis(arylphenoxy)bisphenyls with chlorides of arylenebicarbonic acids in the presence of $AlCl_3$. The polyketones obtained using the most efficient 2,2{}bis(4-bezoylphenoxy) bisphenyl are well soluble in organic solvents and possess high heat resistance.

Also, the aromatic polyketones are produced [334] by the reaction of electrophylic substitution (in dispersion) of copolymer of aliphatic vinyl compound (1-acosen) with N-vinylpirrolydone, at ratio of their links close to equimolar.

The other method of synthesizing aromatic polyketones includes the interaction of monomer of formula $[HC(CN)(NR_2)]Ar$ with 4,4′-difluorobisphenylsulfone in dimethylsulfoxide, dimethylformamide or N-methylpirrolydone at 78–250 °C in the presence of the base [335]. The soluble polyaminonitrile of formula $[C(CN)(NR_2)ArC(CN)(NR_2)C_6H_4SO_2C_6H_4]_n$ is thus produced (Ar denotes m-C_6H_4, NR_2 – stands for group -NCN-$_2CH_2OCH_2CH_2$). When acid hydrolysis of polyaminonitrile is carried out the polysulfoneketone of formula $[COArCOC_6H_4SO_2C_6H_4]_n$ forms with glassing temperature of 192 °C, melting point 257 °C and when 10% mass loss happens at 478 °C. The acid hydrolysis can be performed in the presence of *n*-toluenesulfoacid, trifluoroacetic acid, and mineral acids.

The synthesis of aromatic polyketone particles has been carried out [336] by means of precipitation polycondensation and is carried out at very the low concentration of monomer [0.05 mol/L]. The polyketones are produced from bisphenoxybenzophenone (0.005 mole) or isophthaloylchloride (0.005 mole) in 100 milliliters of 1,2-dichloroethane. Some of obtained particles have highly organized the needle-shaped structure (the whisker crystals). The use of isophthaloyl instead of terephthaloyl at the same low concentration of monomer results in formation of additionally globular particles, the binders of strip structures gives rise. The average size of needle-shaped particles is 1–5 mm in width and 150–250 mm in length.

The synthesis of aromatic high-molecular polyketones by the low-temperature solid-state polycondensation of 4,4′-bisphenoxybenzophenone and isophthaloylchloride in the presence of $AlCl_3$ in 1,2-dichlorethane is possible [337]. Obtained polymers are thermoplasts with glass transition temperature 160 °C and melting point 382 °C.

The polyesterketones are synthesized [338] from dichlorbenzophenone and Na_2CO_3 in the presence of catalyst SiO_2-Cu-salt. The polymer has the negligible number of branched structures and differs from polyesterketones synthesized on the basis of 4-hydroxy-4′-flourobenzophenone on physical properties (its pressed samples have higher crystallinity and orientation).

There exist reports on syntheses of fully aromatic polyketones without single ester bonds [339–349].

The fully aromatic polyketones without ether bonds were produced [339] on the basis of polyaminonitrile, which was synthesized from anions of bis(aminonitrile) and 4,4′{}-difluorobenzophenone using the sodium hydride in mild conditions. The acid hydrolysis of synthesized polyaminonitrile avails one to obtain corresponding polyketone with high thermal properties and tolerance for organic solvents.

They are soluble only in strong acids such as concentrated H_2SO_4. The polyketones have glassing temperature 177–198 °C. Their melting points and temperature of the beginning of decomposition are, respectively 386–500 °C and 493–514 °C [340].

The aromatic polyketones without ester bonds can also be produced by the polymerization of bis(chlorbenzoyl)dimethoxybisphenyls in the presence of nickel compounds [341]. The polymers have the high molecular mass, the amorphous structure, the glass transition temperature 192 °C and 218 °C and form abiding flexible films.

The known aromatic polyketones (the most of them) dissolve in strong acids, or in trifluoroacetic acid mixed with methylenechloride, or in trifluoroacetic acid mixed with chloroform.

It is known that the presence of bulk lateral groups essentially improves the solubility of polyketones, and also improves their thermo-stability. In connection to this, dichloranhydride of 3,3-bis-(4′-carboxyphenyl)phtalide (instead of chloranhydride) was used as initial material for condensation with aromatic hydrocarbons [350]. According to the Friedel-Crafts reaction of electrophylic substitution in variant of low-temperature precipitation polycondensation the high-molecular polyarylenepthalidesterketones have been synthesized. Obtained polymers have greater values of intrinsic viscosity 1.15–1.55 dL/g (in tetrachloroethane). The softening temperatures lies within 172–310 °C, the temperature of the beginning of decomposition is 460 °C. Polyarylenepthalidesterketones dissolve in wide range of organic solvents, form colorless, transparent, abiding (σ = 85–120 MPA) and elastic (ε = 80–300%) films when formed from the melt.

The soluble polyketones – polyarylemethylketones – can be produced [351] by condensation polymerization of 1,4-dihalogenarenes and 1,4-diacetylbenzols in the presence of catalytical palladium complexes, base and phosphoric ligands. The high yield of polymer is seen when using the tetrahydrofurane, o-dichlorbenzole and bisphenyl ester as solvents. The synthesized polymers dissolve in tetrahydrofurane, dichloromethane and hexane, has the decomposition point 357 °C (in nitrogen). It has luminescent properties: emits the green light (490–507 nm) after light irradiation with wavelength of 380 nm.

The Refs. [352–355] report on syntheses of carding aromatic polyketones.

The ridge-like polyarylesterketones have been synthesized by means of one-stage polycondensation in solution of bis(4-nitrophenyl)ketone with phenolphthalein, o-cresolphthaleine, 2,5,2′-5′-tetramethylphenolphthaleine and timolphthaleine [352]. Authors have shown that the free volume within the macromolecule depends on position, type and number of alkyl substitutes.

The homo- and copolyarylenesterketones of various chemical constitution (predominantly, carding ones) have been produced by the reaction of nucleophylic substitution of aromatic activated dihaloid compound [353], and also the "model" homopoyarylenesterketones on the basis of bisphenol, able to crystallize, created. The tendency to crystallization is provided by the combination of fragments of carding bisphenols with segments of hydroquinone (especially, of 4,4′-dihydroxydiphenyl) and increases with elongation of difluoro-derivative (the oligomer homologues of benzophenone), and also in the presence of bisphenyl structure in that fragment. Owing to the presence of carding group, the glass transition temperature of copolymer reaches 250 °C, and the melting point – 300–350 °C.

The crystallizing carding polyarylenesterketones, in difference from amorphous ones, dissolve in organic solvents very badly, they are well soluble in concentrated sulfuric acid at room temperature and when heating to boiling in m-cresol (precipitate on cooling) [354, 355].

There are contributions [356, 357] devoted to the methods of synthesis of aromatic polyketones on the basis of diarylidenecycloalkanes.

The polyketones are produced [356] by the reaction of 2,7-dibenzylidenecyclopentanone (I) and dibenzylideneacetone (II) with dichlorides of different acids (isophthalic, 3,3{}-azodibenzoic and others) in dry chloromethane in the presence of $AlCl_3$. The "model" compounds from I and II and benzoylchloride are also obtained. The synthesized polyketones have intrinsic viscosity 0.36–0.84 dL/g (25 °C, H_2SO_4). They do not dissolve in most organic solvents, dissipate in H_2SO_4. It is determined that the polyketones containing the aromatic links are more stable than those involving aliphatic and azo group. The temperature of 10% and 50% mass loss is 150–250 °C and 270–540 °C for these polyketones.

The polyketones, possessing intrinsic viscosity 0.76–1.18 dL/g and badly dissolving in organic solvents, can be produced by Friedel-Crafts polycondensation of diarylidenecyclopentanone or diarylidenecyclohexanone, chlorides of aromatic or aliphatic diacids, or azodibenzoylchlorides. The temperature of 10% mass loss is 190–300 °C. The in polyketones have the absorption band at wavelength 240–350 nm in ultraviolet spectra (visible range) [357].

The metal-containing polyketones, which do not dissolve in most of organic solvents and easily dissipate in proton dissolvents, have been produced by the reaction of 2,6 [bis(2-ferrocenyl)methylene]cyclohexane with chlorides of dicarboxylic acids [358]. The intrinsic viscosity of metal-containing polyketones is 0.29–0.52 dL/g.

The Refs. [359, 360] report on different syntheses of isomeric aromatic polyketones. Three isomeric aromatic polyketones, containing units of 2-trifluoromethyl- and 2,2{}-dimetoxybisphenylene were synthesized in Ref. [326] by means of direct electrophylic aromatic acylated polycondensation of monomers. Two isomers of polyketone of structure "head-to tail" and "head-to head" contain the links of 2-trifrluoromethyl-4,4′{}-bisphenylene and 2,2{}-dimethoxy-5,5{}-bisphenylene.

There exist several reports on syntheses of polyarylketones, containing bisphthalasinone and methylene [361, 362], naphthalene [363–367] links; containing sulfonic groups [368], carboxyl group in side chain Ref. [369], fluorine [370–372] and on the basis of carbon monoxide and styrol or *n*-sthylstirol [373]. It is shown that methylene and bisphthalasinone links in main chain of polyketones are responsible for its good solubility in m-cresol, chloroform. The links of bisphthalasinone improve thermal property of polymer.

The synthesis of temperature-resistant polyketone has been carried out in Ref. [362] by means of polycondensation of 4-(3-chlor-4-oxyphenyl)-2,3-phthaloasine-1-one with 4,4′-difluorobisphenylketone.

The polymer is soluble in chloroform, N-methylpirrolydone, nitrobenzene and tetrachloroethane, its glass transition temperature is 267 C.

The aromatic polyketones, containing 1,4-naphthalene links were produced in Ref. [363] by the reaction of nucleophylic substitution of 1-chlor-4-(4′-chlorbenzoyl)naphthalene with 1) 1,4-hydroquinone, 2) 4,4′-isopropylidenediphenol, 3) phenolphthalein, 4) 4-(4′-hydroxyphenyl)(2H)-phthalasine-1-one, respectively. All polymers are amorphous and dissipate in some organic solvents. The polymers have good thermo-stability and the high glassing temperatures.

Fluorine containing polyarylketone was synthesized on the basis of 2,3,4,5,6-pentaflourbenzoylbisphenyl esters in Ref. [370]. The polymers possesses good mechanical and dielectric properties, has impact strength, solubility and tolerance for thermooxidative destruction.

The just-produced polyketones are stabilized by treating them with acetic acid solution of inorganic phosphate [371].

The simple fluorine containing polyarylesterketones was obtained in Ref. [372] on the basis of bisphenol AF and 4,4′-difluorobenzophenone. The polymer has the glass transition temperature 163 °C and temperature of 5% mass loss 515 °C, dielectric constant 1.69 at 1 MHz, is soluble well in organic solvents (tetrahydrofurane, dimethylacetamide and others).

The Refs. [374, 375] are devoted to the synthesis of polyesterketones with lateral methyl groups.

The polyarylenesterketones have been produced by the reaction of nucleophylic substitution of 4,4{}-difluorobenzophenone with hydroquinone [374]. The synthesis is held in sulfolane in the presence of anhydrous the K_2CO_3. The increasing content of methylhydroquinone links in polymers leads to increasing of glass transition temperature and lowering of crystallinity degree, melting temperature and activation energy.

The methyl-substituted polyarylesterketones have been produced in Ref. [375] be means of electrophylic polymerization of 4,4{}bis-(0-methylphenoxy)bisphenylketone or 1,4-bis(4-(methylphenoxy)benzoyl)benzene with terephthaloyl or isophthaloyl chloride in 1,2-dichloroethane in the presence of dimethylformamides and $AlCl_3$. The polymers have glassing temperature and melting point 150–170 °C and 175–254 °C.

The simple aromatic polyesterketones, which are of interest as constructional plastics and film materials, capable of operating within the long time at 200 °C, have been synthesized [376–381] by means of polynitrosubstitution reaction of 1,1-dichlor-2, 2-di(4-nitrophenyl)-ethylene and 4,4′-dinitrobenzophenone with aromatic bisphenols.

The simple esterketone oligomer can be synthesized [378] also by polycondensation of aromatic diol with halogen-containing benzophenone at 150–250 °C in organic solvent in the presence of alkali metal compound as catalyst and water.

The particles of polyesterketone have a diameter ≤50 micrometers, intrinsic viscosity 0.5–2 dL/g (35 °C, the ratio *n*-chlorphenol: phenol is 90:10).

The simple polyesterketone containing lateral side cyano-groups, possessing the glass transition temperature in range 161–179 °C, has been produced [379] by low-temperature polycondensation of mix 2,6-phenoxybenzonitrile and 4,4′-bisphenoxybenzophenone with terephthaloylchloride in 1,2-dichloroethane.

The simple polyarylesterketones, containing the carboxyl pendent groups [380] and flexible segments of oxyethylene [381] have been synthesized either.

Russian and foreign scientists work on synthesis of copolymers [382–393] and block copolymers [394–399] of aromatic polyketones.

So, the high-molecular polyketones, having amorphous structure and high values of glassing temperature, have been produced by copolymerizing through the mechanism of aromatic combination of 5,5-bis(4-chlorbemzoyl)-2,2-dimethoxybisphenyl and 5,5-bis(3-chlorbenzoyl)-2,2-dimetoxybishenyl with help of nickel complexes [382].

Terpolymers on the basis of 4,4′-bisphenoxybisphenylsulfone, 4,4′-bisphenoxybenzophenone and terephthaloylchloride were produced in Ref. [383] by low-temperature polycondensation. The reaction was lead in solution of 1,2-dichloroethane in the presence of $AlCl_3$ and N-methyl-2-pyrrolidone. It has been found that with increasing content of links of 4,4′-bisphenoxybisphenylsulfone in copolymer their glassing and dissipation temperatures increase, but melting point and temperature of crystallization decrease.

The new copolyesterketones have been produced in Ref. [384] also from 4,4′-difluorobenzophenone, 2,2{},3,3{},6,6{}-hexaphenyl-4,4′-bisphenyl-1,1{}-diol and hydroquinone by the copolycondensation in solution (sulfolane being the dissolvent) in the presence of bases (Na_2CO_3, K_2CO_3). The synthesized copolyesterketones possess solubility, high thermal stability; have the good breaking strength and good gas-separating ability in relation to CO_2/N_2 and O_2/N_2.

The random copolymers of polyarylesterketones were produced by means of nucleophylic substitution [385] the basis of bisphenol A and carding bisphenols (in particular, phenolphthalein).

The simple copolyesterketones (copolyesterketonearylates) are synthesized [386] and the method for production of copolymers of polyarylates and polycarbonate with polyesterketones is patented [387]. The reaction is held in dipolar aprotone dissolvent in the presence of interface catalyst (hexaalkylguanidinehalogenide).

The statistical copolymers of polyarylesterketones, involving naphthalene cycle in the main chain, can be produced [388] by low-temperature polycondensation of bisphenyloxide, 4,4′-{}-bis(β-naphtoxy)benzophenone with chloranhydrides of aromatic bicarbonic acids – terephthaloylchloride and isophthaloylchloride (I) in the presence of catalytical system $AlCl_3$/N–methylpirrolydone/$ClCH_2CH_2Cl$ (copolymers are characterized by improved thermo- and chemical stability), and also by the reaction of hydroquinone with 1,4-bis(4,4′-flourobenzoyl)naphthalene (II) in the presence of sodium and potassium carbonates in bisphenylsulfone [389].

The glass transition temperature of polyarylesterketones is going up, and melting point and temperature of the beginning of the destructions are down with increasing concentration of links of 1,4-naphthalene in main chain of copolymers.

The polyarylesterketone copolymers, containing lateral cyano-groups, can be synthesized [390] on the basis of bisphenyl oxide, 2,6-bisphenoxybenzonitrile (I) and terephthaloylchloride in the presence of $AlCl_3$, employing 1,2-dichloroethane as the dissolvent, N-mthyl-2-pyrrolidon as the Lewis base. With increasing concentration of links of I the crystallinity degree and the melting point of copolymers decrease while the glass transition temperature increase. The temperature of 5%of copolymers mass loss is more than 514 °C (N_2).

Copolymers, containing 30–40 weight % of links of I, possess higher therm0-stability (350 ± 10 °C), good tolerance for action of alkalis, bases and organic solvents.

The copolymers of polyarylesterketones, containing lateral methyl groups, can be produced by low-temperature polycondensation of 2,2{}-dimethyl-4,4{}-bisphenoxybisphenyleneketone (I) or 1,4-[4-(2-methylphenoxy)benzoyl]benzene (II) and bisphenylester, terephthaloylchloride [391]. The synthesis is held in 1,2-dichloroethane in the presence of dimethylformamides and $AlCl_3$ as catalyst. With increasing concentration of links of I or II in copolymers the glass transition temperature increases, and the melting point and the crystallinity degree decreases.

The copolymers of polyarylketones, containing units of naphthalenesulfonic acid in lateral links, are use in manufacturing of proton-exchange membranes [392].

The polymeric system, suitable as proton-exchanged membrane in fuel cells, was developed in Ref. [393] on the basis of polyarylenesterketones.

The polyarylesterketones are firstly treated with metasulfonic acid within 12 h at 45 C up to sulfur content of 1,2%, and then sulfonize with oleum at 45 C up to sulfur content 5% (the degree of sulfonation 51%).

The high-molecular block-copolyarylenesterketones have been synthesized on the basis of 4,4′-difluorobenzophenone and number of bisphenols by means of reaction of nucleophylic substitution of activated arylhalogenide in dimethylacetamide (in the presence of potassium carbonate) [394, 395]. It has been identified the cutback of molecular weight of polymer when using bisphenolate of 4,4′-(isopropylidene)bisphenol.

The block-copolyesterketones are synthesized [396] on the basis of dichloranhydride-1,1′-dichlor-2,2′-bis(*n*-carboxyphenyl)ethylene and various dioxy-compounds with complex of worthy properties. It is possible to create, on the basis of obtained polymers, the constructive and film materials possessing lower combustibility and high insulating properties. Some synthesized materials are promising as modifiers of high-density polyethylene.

The block copolymers, containing units of polyarylesterketones and segments of thermotropic liquid-crystal polyesters of different length, are synthesized by means of high-temperature polycondensation in solution

[397, 398]; the kinetics, thermal and liquid-crystal properties of block-copolymers have been studied.

The polyarylenesterketones have been obtained [399], which contain blocks of polyarylesterketones, block of triphenylphosphite oxide and those of binding agent into the cycloaliphatic structures. The film materials on their basis are used as polymer binder in thermal control coatings.

The Refs. [400–402] informed on production of polyester-α-diketones using different initial compounds.

Non-asymmetrical polyester-α-diketones are synthesized [400] on the basis of 4-flouru-4′(*n*-fluorophenylglyoxalyl)benzophenone. Obtained polymers are amorphous materials with glassing temperature 162–235 °C and have the temperatures of 10% mass loss in range of 462–523 °C.

The polyester-α-diketones synthesized in Ref. [401] on the basis of 2,2-bis[4-(4-fluorophenylglyoxalyl)phenyl]hexafluoropropane are soluble polymers. They exhibit the solubility in dimethylformamide, dimethylacetamide, N-methylpirrolydone, tetrahydrofurane and chloroform. The values of glassing temperature lie in interval 182–216 °C, and temperatures of 10% mass loss are within 485–536 °C and 534–556 °C in air and argon, respectively.

The number of new polyester-α-diketones in the form of homopolymers and copolymers has been produced [402]. Polyester-α-diketones have the amorphous structure, include α-diketone, α-hydroxyketone and ester groups, they are soluble in organic solvents. The films with lengthening of 5–87%, solidity and module under tension, respectively 54–83 MPa and 1.6–3 GPa are produced by casting from chloroform solution.

The quinoxalline polymers, possessing higher glassing and softening temperatures and better mechanical properties compared to initial polymers, have been produced by means of interaction of polyester-α-diketones with α-phenylenediamine at 23 °C in m-cresol. Polyquinoxalines are amorphous polymers, soluble in chlorinated, amide and phenol dissolvents with η_{lim} = 0.4–0.6 dL/g (25 °C, in N-methylpirrolydone of 0.5 gram/deciliter).

It is reported [403] on liquid-crystal polyenamineketone and on mix of liquid-crystal thermotropic copolyester with polyketone [404]. The mix is prepared by melts mixing. It has been found that the blending agents are partially compatible; the mixes reveal two glass transition temperatures.

The transitions of polyketones into complex polyesters are possible.

So, when conducting the reaction at 65–85 °C in water, alcohol, ester, acetonitrile, the polyketone particles of size 0.01–100 micrometers transform the polyesters by the oxidation under the peroxide agents: peroxybenzoic, m-chloroperoxybenzoic, peroxyacetic, triflouruperoxyacetic, monoperoxyphthalic, monoperoxymaleine acids, combinations of H_2O_2 and urea or arsenic acid [405, 406].

For increasing of basic physical-mechanical characteristics and reprocessing, in particular of solubility, the synthesis of aromatic polyketones is lead through the stages of formation of oligomers with end functional groups accompanied by the production of block-copolyketones or through the one-stage copolycondensation initial monomers with production of copolyketones.

So, unsaturated oligeketons are produced [407, 408] by the condensation of aromatic esters (e.g., bisphenyl ester, 4,4′-bisphenoxybisphenyl ester and others) with maleic anhydride in the presence catalyst $AlCl_3$.

The oligoketones with end amino groups can be produced [409] on the basis of dichloranhydride of aromatic dicarboxylic acid, aromatic carbohydrate and telogen (N-acylanylode) according Friedel-Crafts reaction in organic dissolvent.

The oligoketones can be produced [410] by polycondensation of bisphenyloxide and 4-fluorobenzoylchloride in solution, in the presence of $AlCl_3$. It has been found that the structure of synthesized oligomer is crystalline.

The synthesis of oligoketones, containing phthaloyl links, can be implemented [411] by the Friedel-Craft reaction of acylation.

The cyclical oligomers of phenolphthalein of polyarylenestersulfonketone can be produced [412] by cyclical depolymerization of corresponding polymers in dipolaraprotic solvent (dimethylformamide), dimethylacetate in the presence of CsF as catalyst.

The aromatic oligoesterketones can be produced by means of interaction between 4,4′-dichlorbisphenylketone and 1,1-dichlor-2,2-di(3,5-dibrom-*n*-oxyphenyl)ethylene in aprotic dipolar dissolvent (dimethylsulfoxide) at 140 °C in inert gas [413]. The copolyesterketones of increased thermo-stability, heat- and fire-resistance can be synthesized on the basis of obtained oligomers.

The works on production of oligoketones and synthesis of aromatic polyketones on their basis are held in Kh.M. Berbekov Kabardino-Balkarian State University.

So, availing the method of high-temperature polycondensation in environment of dimethylsulfoxide, the oligoketones are produced with edge hydroxyl groups with the degree of condensation 1–20 from bisphenols (diane or phenolphthalein) and dichlorbenzophenone. The condensation on the second stage is carried out at room temperature in 1,2-dichloroethane in the presence of HCl as acceptor and triethylamine as catalyst with introduction of the diacyldichloride of 1,1-dichlor-2,2-(*n*-carboxyphenyl) ethylene into the reaction. The reduced viscosity of copolyketones lies within 0.78–3.9 dL/g for polymers on the basis of diane oligoketones and 0.50–0.85 dL/g for polymers on the basis of phenolphthalein oligoketones. The mix of the tetrachloroethane and phenol in molar ratio 1:1 at 23 °C was taken as a solvent [414].

The aromatic block-copolyketones have been synthesized by means of polycondensation of oligoketones of different composition and structure with end OH-groups on the basis of 4,4′-dichlorbenzophenone and dichloranhydrides mixed with iso- and terephthalic acids in environment of 1,2-dichloroethane [415–418].

The synthesis is lead via the method of acceptor-catalyst polycondensation. The double excess of triethylamine in respect to oligoketones is used as acceptor-catalyst. Obtained block-copolyketones possess good solubility in chlorinated organic solvents and can be used as temperature-resistant high-strength durable constructional materials.

7.7 AROMATIC POLYESTERSULFONE KETONES

Along with widely used polymer materials of constructional assignment such as polysulfones, polyarylates, polyaryleneketones, etc, the number of researches has recently, within the last decades, appeared which deal with production and study of properties of polyestersulfoneketones [290, 419–433]. The advantage of given polymeric materials are that these simultaneously combine properties of both polysulfones and polyaryleneketones and it allows one to exclude some or the other disadvantages of two classes of polymers. It is known, that high concentration of ketone groups results in greater T_{glass} and T_{flow}, respectively, or – in better heat stability. On the other hand, greater concentration of ketone groups results in worsening of reprocessibility of polyarylenesterketones. That is why the sulfonation of the latter has been performed. The solubility increases with increasing of the degree of sulfonation. Polyestersulfoneketones gain solubility in dichloroethane, chloroform, dimethylformamide. The combination of elementary links of polysulfone with elementary links of polyesterketone also improves the fluidity of composition during extrusion.

The Refs. [424–433] report on various methods of production of polyestersulfoneketones.

The 4,4{}-bis(phenoxy)bisphenylsulfone has been synthesized by means of reaction of phenol with bis-(4-chlorphenyl)sulfone and the low-temperature polycondensation of the product has been carried out in the melt with tere- and isophthaloylchloride resulting in the formation of polyarylenestersulfonesterketoneketones. The polycondensation is conducted in 1,2-dichlorethane at presence of $AlCl_3$ and N-methylpirrolydone. Compared to polyesterketoneketones synthesized polymers have greater temperatures of glassing and decomposition and lower melting point and also possess higher thermal resistance [424].

The block-copolymers have been created by means of polycondensation of low-molecular polyesterketone containing the remnants of chloranhydride of carbonic acid and 4,4{}-bisphenoxybisphenylsulfone (I) as end groups. The increase in concentration of component I result in greater glassing temperature and lower melting point of block-copolymers.

The block-copolymers containing 32.63–40.7% of component I have glassing temperature and melting point, respectively 185–193 °C and 322–346 °C, durability and modulus in tension, respectively 86.6–84.2 MPa and 3.1–3.4 GPa and tensile elongation 18.5–20.3%.

Block-copolymers have good thermal properties and could be easily reprocessed in melt [425].

Polyestersulfoneketones have been produced be means of electrophylic Friedel-Crafts acylation in the presence of dimethylformamides and anhydrous $AlCl_3$ in environment of 1,2-dichlorethane on the basis of simple 2- and 3-methylbisphenyl esters and 4,4{}-bis(4-chloroformylpheoxy) bisphenylsulfone. The copolymer has the molecular weight of 57,000–71,000, its temperatures of glassing and decomposition are 160.5–167,0 C and > 450 °C, respectively, the coke end is 52–57% (N_2). Copolyestersulfoneketones are well soluble in chloroform and polar dissolvents (dimethylformamide and others) and form transparent and elastic films [426].

Some researches of Chinese researchers are devoted to the synthesis and study of properties of copolymers of arylenestersulfones and esterketones [427–429].

The block-copolyesters based on oligoesterketones and oligoestersulfones of various degree of condensation have been produced in Ref. [430] by means of acceptor-catalyst polycondensation; the products contain simple and complex ester bonds.

Polyestersulfoneketones possessing lower glassing temperatures and excellent solubility have been synthesized in Ref. [431] by means of low-temperature polycondensation of 2,2{}, 6,6{}-tetramethylbisphenoxybisphenylsulfone, iso- and terephthaloylchloride in the presence of $AlCl_3$ and N-methylpirrolydone in environment of 1,2-dichlorethane. The increasing concentration of iso- and terephthaloylchloride results in greater glassing temperatures of statistical copolymers.

The manufacture of ultrafiltration membranes is possible on the basis of mix polysulfone-polyesterketone [432, 433]. Transparent membranes obtained by means of cast from solution have good mechanical properties in both dry and hydrated state and keep analogous mechanical properties after exposition in water (for 24 h at 80 °C). The maximal conductivity of membranes at 23 °C is 4.2×10^{-2} Sm/cm while it increases to 0.11 Sm/cm at 80 °C.

In spite of a number of investigations in the area of synthesis and characterization of polyestersulfoneketones, there are no data in literature on the polyestersulfoneketones of block-composition based on bisphenylolpropane (or phenolphthalein), dichlorbisphenylsulfone and dichlorbenzophenone, terephthaloyl-bis(*n*-oxybenzoylchloride).

Russian references totally lack any researches on polyestersulfoneketones, not speaking on the production of such polymers in our country. Accounting for this, we have studied the regularities of the synthesis and produced block-copolyestersulfoneketones of some valuable properties [423]. The main structure elements of the polymers are rigid and extremely thermo-stable phenylene groups and flexible, providing for the thermoplastic reprocessibility, ester, sulfone and isopropylidene bridges.

Synthesized, within given investigation, polyestersulfoneketones on the basis of phthalic acid; polyarylates on the basis of 3,5-dibromine-*n*-oxybenzoic acid and phthalic acids and copolyesters on the basis of terephthaloyl-bis(*n*-oxybenzoic) acid are of interest as heat-resistant and film materials which can find application in electronic, radioelectronic, avia, automobile, chemical industries and electotechnique as thermo-stable constructional and electroisolation materials as well as for the protection of the equipment and devices from the influence of aggressive media.

KEYWORDS

- **Monomers**
- **N-oxybenzoic acid**
- **Oligoesters**
- **Polycondensation**
- **Polyesters**
- **Synthesis**
- **Tere- and isophthalic acids**

REFERENCES

1. Sokolov, L. V. (1976). *The Synthesis of Polymers: Polycondensation Method.* Moscow: Himia, 332 p. [in Russian].
2. Morgan, P. U. (1970). *Polycondensation Processes of Polymers Synthesis.* Leningrad: Himia, 448 p. [In Russian].
3. Korshak, V. V., & Vinogradova, S. V. (1968). *Equilibrium Polycondensation.* Moscow: Himia, 441 p. [in Russian].
4. Korshak, V. V., & Vinogradova, S. V. (1972). *Non-Equilibrium Polycondensation.* Moscow: Nauka, 696 p. [in Russian].
5. Sokolov, L. B. (1979). *The Grounds for the Polymers Synthesis by Polycondensation Method.* Moscow: Himia, 264 p. [in Russian].
6. Korshak, V. V., & Kozyreva, N. M. (1979). *Uspehi Himii, 48(1),* 5–29.
7. (1977). *Encyclopedia of Polymers.* Moscow: Sovetskaia Enciklopedia, *3,* 126–138. [in Russian].
8. Korshak, V. V., & Vinogradova, S. V. (1964). *Polyarylates.* Moscow: Nauka, 68 p. [in Russian].
9. Askadskii, A. A. (1969). *Physico-Chemistry of Polyarylates.* Moscow: Himia, 211 p. [in Russian].
10. Korshak, V. V., & Vinogradova, S. V. (1958). *Heterochain Polyesters.* Moscow: Academy of Sciences of USSR, 403 c. [in Russian].
11. Lee, G., Stoffi, D., & Neville, K. (1972). *Novel Linear Polymers.* Moscow: Himia, 280 p. [in Russian].
12. Sukhareva, L. A. (1987). *Polyester Covers: Structure and Properties.* Moscow: Himia, 192 p. [in Russian].
13. Didrusco, G., & Valvaszori, A. (1982). Prospettive nel Campo Bei Tecnopolimeri Tecnopolime Resine, *(5),* 27–30.
14. Abramov, V. V., Zharkova, N. G., & Baranova, N. S. (1984). Abstracts of the All-Union Conference "Exploiting Properties of Constructional Polymer Materials". Nalchik, 5. [in Russian].

15. Tebbat Tom. (1975). Engineering Plastics: Wonder Materials of Expensive Polymer Plauthings *Eur. Chem. News*, *27,* 707.
16. Stoenesou, F. A. (1981). Tehnopolimeri. *Rev. Chem*, *32(8),* 735–759.
17. Nevskii, L. B., Gerasimov, V. D., & Naumov, V. S. (1984). In Abstracts of the All-Union Conference "Exploiting Properties of Constructional Polymer Materials". Nalchik, 3. [in Russian]
18. Mori, Hisao. (1975). *Jap. Plast*, *26(8),* 23–29.
19. Karis, T., Siemens, R., Volksen, W., & Economy, J. (1987). Melt Processing of the Phbahomopolymer in Abstracts of the 194-th ACS National Meeting of American Chemical Society. New Orleans: Los Angeles, August 30–September 4, Washington DC, *1987*, 335–336.
20. Buller, K. (1984). *Heat-and Thermostable Polymers*. Moscow: Himia, 343 p. [in Russian].
21. Crossland, B., Knight, G., & Wright, W. (1986). The Thermal Degradation of Some Polymers Based upon P-Hydroxybenzoic Acid Brit. *Polym. J*, *18(6),* 371–375.
22. George, E., & Porter, R. (1988). Depression of the Crystalnematic Phase Transition in Thermotropic Liquid Crystal Copolyesters. *J. Polym. Sci.*, 26*(1),* 83–90.
23. Yoshimura, T., & Nakamura, M. (1986). Wholly Aromatic Polyester US Patent 4609720. Publ. 02.09.86.
24. Ueno, S., Sugimoto, H., & Haiacu, K. (1984). Method for Producing Polyarylates Japan Patent Application 59–120626. Publ. 12.07.84.
25. Ueno, S., Sugimoto, H., & Haiacu, K. (1984). Method for Producing Aromatic Polyesters. Japan Patent Application 59–207924. Publ. 26.11.84.
26. Yu Michael, C. Polyarylate Formation by Ester Interchange Reaction Using–Gamma Lactones as Diluent US Patent 4533720.
27. Higashi, F., & Mashimo, T. (1986). Direct Polycondensation of Hydroxybenzoic Acids with Thionylchloride in Pyridine. *J. Polym. Sci.: Polym. Chem. Ed*, *24(7),* 177–1720.
28. Bykov, V. V., Tyuneva, G. A., Trufanov, A. N., et al. (1986). *Izvestia Vuzov. Khim. i Him. Technol*, *29(12),* C. 20–22.
29. Tedzaki K Hiroaka. (1973). Method for Producing Polymers of Oxybenzoic Acid. Japan Patent 48–23677. Publ. 16.07.73.
30. Process for Preparation of Oxybenzoyl Polymer US Patent 3790528. Publ. 05.02.74.
31. Sima, Takeo, Yamasiro, Saiti, & Inada, Hiroo. (1973). Method for Producing Polyesters of *n*-oxybenzoic Acid. Japan Patent 48–37–37355. Publ. 10.11.73.
32. Sakano, Tsutomu, & Miesi, Takehiro. (1984). Polyester Fiber Japan Patent Application 59–199815. Publ. 13.11.84.
33. Higashi, Fukuji, & Yamada, Yukiharu. Direct Polycondensation of Hydroxybenzoic Acids with Bisphenyl Chlorophosphate in the Presence of Esters .*J. Polym. Sci.: Polym. Chem. Ed.*
34. Adxuma, Fukudzi. (1985). Method for Producing Complex Polyesters. Japan Patent Application 60–60133. Publ. 06.04.85.
35. Chivers, R. A., Blackwell, J., & Gutierrez, G. A. (1985). X-ray Studies of the Structure of HBA/HNA Copolyesters in Proceedings of the 2-nd Symposium "*Div. Polym. Chem. Polym. Liq. Cryst.*" Washington DC, New York, London, 153–166.

36. Sugijama, H., Lewis, D., & White, J. Structural Characteristics, Rheological Properties, Extrusion and Melt Spinning of 60/40 Poly(Hydroxybenzoic Acidcoethylene Terephtalate).
37. Blackwell, J., Dutierrez, G., & Chivers, R. (1985). X-ray Studies of Thermotropic Copolyesters in Proceedings of the 2nd Symposium "*Div. Polym. Chem. Polym. Liq. Cryst.*" Washington DC, New York, London, 167–181.
38. Windle, A., Viney, C., & Golombok, R. (1985). Molecular Correlation in Thermotropic Copolyesters Faraday Discuss. *Chem. Soc*, *79*, 55–72.
39. Calundann Gordon, & Meet, W. Processable Thermotropic Wholly Aromatic Polyester Containing Polybenzoyl Units US Patent 4067852.
40. Morinaga Den, Inada Hiroo, & Kuratsudzi Takatozi. (1977). Method for Producing Thermostable Aromatic Polyesters Japan Patent Application 52–121626. Publ. 13.10.77.
41. Sugimoto Hiroaki, & Hanabata Makoto. (1983). Method for Producing Aromatic Copolyesters Japan Patent Application 58–40317. Publ. 09.03.83.
42. Dicke Hans-Rudolf, & Kauth Hermann. Thermotrope Aromatische Polyester Mit Hoher Steifigkeit, Verfahren Zu Ihrer Herstellung Und Ihre Verwenaung Zur Herstellung Ven Formkorpern, Filamenten, Fasern und Folien Germany Patent Application 3427886.
43. Cottis Steve, G. Production of Thermally Stabilized Aromatic Polyesters US Patent 4639504.
44. Matsumoto Tetsuo, Imamura Takayuki, & Kagawa Kipdzi. (1987). Polyester Fiber. Japan Patent Application 62–133113. Publ. 1987.
45. Ueno Ryudzo, Masada Kachuiasu, & Hamadzaki Yasuhira. (1987). Complex Polyesters. Japan Patent Application 62–68813. Publ. 28.03.87.
46. Tsai, Hond–Bing, Lee Chyun, & Chang, Nien-Shi. (1990). Effect of Annealing on the Thermal Properties of Poly (4-Hydrohybenzoate-Co-Phenylene Isophthalates) Macromol. *Chem*, *191(6)*, 1301–1309.
47. Paul, K. T. (1986). Fire Resistance of Synthetic Furniture. Detection Methods Fire and Materials, *10(1)*, 29–39.
48. Iosida Tamakiho, & Aoki Iosihisa. (1986). Fire Resistant Polyester Composition. Japan Patent Application 61–215645. Publ. 25.09.86.
49. Wang, Y., Wu, D. C., Xie, X. G., & Li, R. X. (1996). Characterization of Copoly(p-Hydroxybenzoate/Bisphenol-A Terephthalate) by NMR-Spectroscopy. *Polym. J*, *28(10)*, 896–900.
50. Yerlikaya Zekeriya, Aksoy Serpil, & Bayramli Erdal. (2001). Synthesis and Characterization of Fully Aromatic Thermotropic Liquid-Crystalline Copolyesters Containing m-hydroxybenzoic Acid Units *J. Polym. Sci. A*, *39(19)*, 3263–3277.
51. Pazzagli Federico, Paci Massimo, Magagnini Pierluigi, Pedretti Ugo., Corno Carlo, Bertolini Guglielmo, & Veracini Carlo, A. (2000). Effect of Polymerization Conditions on the Microstructure of a Liquid Crystalline Copolyester. *J. Appl. Polym. Sci.*, *77(1)*, 141–150.
52. Aromatic Liquid-Crystalline Polyester Solution Composition US Patent 6838546. International Patent Catalogue C 08 J 3/11, C 08 G 63/19, 2005.

53. Yerlikaya Zekeriya, Aksoy Serpil, & Bayramli Erdal. (2002). Synthesis and Melt Spinning of Fully Aromatic Thermotropic Liquid Crystalline Copolyesters Containing m-hydroxybenzoic Acid Units. *J. Appl. Polym. Sci., 85(12),* 2580–2587.
54. Wang Yu-Zhang, Cheng Xiao-Ting, & Tang Xu-Dong. (2002). Synthesis, Characterization, and Thermal Properties of Phosphorus-Containing, Wholly Aromatic Thermotropic Copolyesters. *J. Appl. Polym. Sci., 86(5),* 1278–1284.
55. Liquid-Crystalline Polyester Production Method US Patent 7005497. International Patent Catalogue C 08 G 63/00, 2006.
56. Wang Jiu-fen, Zhang Na., & Li Cheng-Jie. (2005). Synthesis and Study of Thermotropic Liquid-Crystalline Copolyester. PABA.ABPA.TPA. *Polym. Mater. Sci. Technol, 21(1),* 129–132.
57. (2001). Method of Producing Thermotropic Liquid Crystalline Copolyester, Thermotropic Liquid Crystalline Copolyester Composition Obtained by the Same Composition US Patent 6268419. International Patent Catalogue C 08 K 5/51.
58. Hsiue Lin-tee, Ma Chen-chi M., & Tsai Hong-Bing. (1995). Preparation and Characterizations of Thermotropic Copolyesters of p-hydroxybenzoic Acid, Sebacic Acid, and Hydroquinone. *J. Appl. Polym. Sci., 56(4),* 471–476.
59. Frich Dan., Goranov Konstantin, Schneggenburger Lizabeth, & Economy James. (1996). Novel High-Temperature Aromatic Copolyester Thermosets: Synthesis, Characterization, and Physical Properties Macromolecules, *29(24),* 7734–7739.
60. Dong Dewen, Ni Yushan, & Chi Zhenguo. (1996). Synthesis and Properties of Thermotropic Liquid-Crystalline Copolyesters Containing Bis-(4-oxyphenyl)Methanone. II. Copolyesters from Bis-(4-oxyphenyl)Methanone, Terephthalic Acid, n-oxybenzoic Acid and Resorcene. *Acta Polym. (2)*, 153–158.
61. Teoh, M. M., Liu, S. L., & Chung, T. S. (2005). Effect of Pyridazine Structure on Thin-Film Polymerization and Phase Behavior of Thermotropic Liquid Crystalline Copolyesters. *J. Polym. Sci. B*, *43(16),* 2230–2242.
62. Liquid Crystalline Polyesters having a Surprisingly Good Combination of a Low Melting Point, a High Heat Distortion Temperature, a Low Melt Viscosity, and a High Tensile Elongation US Patent 5969083. International Patent Catalogue C 08 G 63/00. 1999
63. Aromatic complex polyester US Patent 6890988. International Patent Catalogue C 08 L 5/3477. 2005.
64. Process for Producing Amorphous Anisotropic Melt-Forming Polymers Having a High Degree of Stretchability and Polymers Produced by Same US Patent 6207790. *International Patent Catalogue* C 08 G 63/00. 2001.
65. Process for Producing Amorphous Anisotropic Melt-Forming Polymers Having a High Degree of Stretchability and Polymers Produced by Same US Patent 6132884. *International Patent Catalogue* B 32 B 27/06. 2000.
66. Process for Producing Amorphous Anisotropic Melt-Forming Polymers having a High Degree of Stretchability US Patent 6222000. *International Patent Catalogue* C 08 G 63/00. 2001.
67. He Chaobin, Lu Zhihua, Zhao Lun., & Chung Tai-Shung. (2001). Synthesis and Structure of Wholly Aromatic Liquid-Crystalline Polyesters Containing Meta-and Ortho-linkages. *J. Polym. Sci. A, 39(8),* 1242–1248.

68. Choi Woon-Seop, Padias Anne Buyle, & Hall, H. K. (2000). LCP Aromatic Polyesters by Esterolysis Melt Polymerization. *J. Polym. Sci. A, 38(19),* 3586–3595.
69. Chung Tai-Shung, & Cheng Si-Xue. (2000). Effect of Catalysts on Thin-Film Polymerization of Thermotropic Liquid Crystalline Copolyester. *J. Polym. Sci. A, 38(8),* 1257–1269.
70. Collins, T. L. D., Davies, G. R., & Ward, I. M. (2001). The Study of Dielectric Relaxation in Ternary Wholly Aromatic Polyesters Polym. *Adv. Technol, 12(9),* 544–551.
71. Shinn Ted-Hong, Lin Chen-Chong, & Lin David, C. (1995). Studies on Co [Poly(Ethylene Terephthalate-p-oxybenzoate)] Thermotropic Copolyester: Sequence Distribution Evaluated from TSC Measurements. *Polym, 36(2),* 283–289.
72. Poli Giovanna, Paci Massimo, Magagnini Pierluigi, Schaffaro Roberto, & La Mantia Francesco, P. (1996). On the use of PET-LCP Copolymers as Compatibilizers for PET/ LCP Blends. *Polym. Eng. and Sci., 36(9),* 1244–1255.
73. Chen Yanming. (1998). The Study of Liquid-Crystalline Copolyesters PHB/PBT, Modified with HQ-TRA. *J. Fushun Petrol. Inst, 18(1),* 26–29.
74. Wang Jiu-fen, Zhu Long-Xin., & Huo Hong-Xing. (2003). The Method for Producing Thermotropic Liquid-Crystalline Complex Copolyester of Polyethyleneterephthalate. *J. Funct. Polym, 16(2),* 233–237.
75. Liu Yongjian, Jin Yi., Bu Haishan, Luise Robert, R., & Bu Jenny. (2001). Quick Crystallization of Liquid-Crystalline Copolyesters Based on Polyethyleneterephthalate. *J. Appl. Polym. Sci., 79(3),* 497–503.
76. Flores, A., Ania, F., & Balta Calleja, F. J. (1997). Novel Aspects of Microstructure of Liquid Crystalline Copolyesters as Studied by Microhardness: Influence of Composition and Temperature. *Polym, 38(21),* 5447–5453.
77. Hall, H. K.(Jr)., Somogyi Arpad, Bojkova Nina, Padias Anne, B., & Elandaloussi El Hadj. (2003). MALDI-TOF Analysis of all-Aromatic Polyesters / in PMSE Preprints. Papers Presented at the Meeting of the Division of Polymeric Materials Science and Engineering of the American Chemical Society (New Orleans, 2003). *Amer. Chem. Soc, 88,* 139–140.
78. Takahashi Toshisada, Shoji Hirotoshi, Tsuji Masaharu, Sakurai Kensuke, Sano Hirofumi, & Xiao Changfa. (2000). The Structure and Stretchability in Axial Direction of Fibers from Mixes of Liquid-Crystalline all-Aromatic Copolyesters with Polyethyleneterephthalate. *Fiber, 56(3),* 135–144.
79. Watanabe Junji, Yuaing Liu., Tuchiya Hitoshi, & Takezoe Hideo. (2000). Polar Liquid Crystals Formed from Polar Rigid-Rod Polyester Based on Hydroxybenzoic Acid and Hydroxynaphthoic Acid. *Mol. Cryst. Liq. Cryst. Sci. Technol.* A, *346,* 9–18.
80. Juttner, G., Menning, G., & Nguyen, T. N. (2000). Elastizitatsmodul Und Schichtenmorphologie Von Spritzgegossenen LCP-Platten. *Kautsch. and Gummi. Kunstst, B. 53.* S. 408–414.
81. Bharadwaj Rishikesh, & Boyd Richard H. (1999). Chain Dynamics in the Nematic Melt of an Aromatic Liquid Crystalline Copolyester: A Molecular Dynamics Simulation Study. *J. Chem. Phys, 1(20),* 10203–10211.
82. Wang, Y., Wu, D. C., Xie, X. G., & Li, R. X. (1996). Characterization of Copoly(p-Hydroxybenzoate/Bisphenol-A terephthalate) by NMR-Spectroscopy *Polym. J, 28(10),* 896–900.

83. Ishaq, M., Blackwell, J., & Chvalun, S. N. (1996). Molecular Modeling of the Structure of the Copolyester Prepared from p-hydroxybenzoic Acid, Bisphenol and Terephthalic Acid. *Polym, 37(10),* 1765–1774.
84. Cantrell, G. R., McDowell, C. C., Freeman, B. D., & Noel, C. (1999). The Influence of Annealing on Thermal Transitions in a Nematic Copolyester. *J. Polym. Sci. B, 37(6),* 505–522.
85. Bi Shuguang, Zhang, Yi., Bu Haishan, Luise Robert, R., & Bu Jenny, Z. (1999). Thermal Transition of a Wholly Aromatic Thermotropic Liquid Crystalline Copolyester. *J. Polym. Sci. A, 37(20),* 3763–3769.
86. Dreval, V. E., Al-Itavi, Kh. I., Kuleznev, V. N., Bondarenko, G. N., & Shklyaruk, B. F. (2004). *Vysokomol. Soed. A, 46(9),* 1519–1526.
87. Tereshin, A. K., Vasilieva, O. V., Avdeev, N. N., Bondarenko, G. N., & Kulichihin, V. G. (2000). *Vysokomol. Soed. A-B, 42(6),* 1009–1015.
88. Yamato Masafumi, Murohashi Ritsuko, Kimura Tsunehisa, & Ito Eiko. (1997). Dielectric β-Relaxation in Copolymer Ethyleneterephthalate-p-hydroxybenzoic Acid. *J. Polym. Sci. Technol, 54(9),* 544–551.
89. Carius Hans-Eckart, Schonhals Andreas, Guigner Delphine, Sterzynski Tomasz, & Brostow Witold. (1996). Dielectric and Mechanical Relaxation in the Blends of a Polymer Liquid Crystal with Polycarbonate. *Macromolecules, 29(14),* 5017–5025.
90. Tereshin, A. K., Vasilieva, O. V., Bondarenko, G. N., & Kulichihin, V. G. (1995). *Influence of Interface Interaction on Rheological Behavior of Mixes of Polyethyleneterephthalate with Liquid-Crystalline Polyester.* In Abstracts of the III Russian Symposium on Liquid-Crystal Polymers. Chernogolovka, 124. [in Russian].
91. Dreval, V. E., Kulichihin, V. G., Frenkin, E. I., & Al-Itavi, Kh. I. (2000). *Vysokomol. Soed. A-B, 42(1),* 64–70.
92. Kotomin, S. V., & Kulichihin, V. G. (1996). *Determination of the Flow Limit of LQ Polyesters with Help of Method of Parallel-Plate Compression.* In Abstracts of the 18-th Symposium of Rheology. Karacharovo, 61. [in Russian].
93. Zhang Guangli, Yan Fengqi, Li Yong, Wang Zhen, Pan Jingqi, & Zhang Hongzhi. (1996). The Study of Liquid-Crystalline Copolyesters of *n*-oxybenzoic Acid and Polyethyleneterephthalate. *Acta Polym. Sin, (1),* 77–81.
94. Dreval, V. E., Frenkin, E. I., Al-Itavi, Kh. I., & Kotova, E. V. (1999). Some Thermophysical Characteristics of Liquid-Crystalline Copolyester Based on Oxybenzoic Acid and Polyethyleneterephthalate at High Pressures. In Abstracts of the IV-th Russian Symposium (involving international participants) “Liquid Crystal Polymers”. Moscow, 62. [in Russian].
95. Brostow Witold, Faitelson Elena, A., Kamensky Mihail, G., Korkhov Vadim, P., & Rodin Yuriy, P. (1999). Orientation of a Longitudinal Polymer Liquid Crystal in a Constant Magnetic Field. *Polym, 40(6),* 1441–1449.
96. Dreval, V. E., Hayretdinov, F. N., Kerber, M. L., & Kulichihin, V. G. (1998). *Vysokomol. Soed. A-B, 40(5),* 853–859.
97. Al-Itavi, Kh. I., Frenkin, E. I., Kotova, E. V., Bondarenko, G. N., Shklyaruk, B. F., Kuleznev, V. N., Dreval, V. E., & Antipov, E. M. (2000). *Influence of High Pressure on Structure and Thermophysical Properties of Mixes of Polyethyleneterephthalate with Liquid-Crystalline Polymer.* In Abstracts of the 2-nd Russian Kargin Symposium

Chemistry and Physics of Polymers in the Beginning of the 21 Century. Chernogolovka, Part 1. P. 1/13. [in Russian].
98. Garbarczyk, J., & Kamyszek, G. (2000). Influence of Magnetic and Electric Field on the Structure of IPP in Blends with Liquid Crystalline Polymers. In Abstracts of the 38-th Macromolecular IUPAK Symposium. Warsaw, *3,* 1195.
99. Dreval, V. E., Frenkin, E. I., & Kotova, E. D. (1996). Dependence of the Volume Form the Temperature and Pressure for Thermotropic LQ-Polymers and their Mixes with Polypropylene. In Abstracts of the 18-th Symposium on Rheology. Karacharovo, 45. [in Russian].
100. Al-Itavi, Kh. I., Dreval, V. E., Kuleznev, V. N., Kotova, E. V., & Frenkin, E. I. (2003). *Vysokomol. Soed, 45(4),* 641–648.
101. Plotnikova, E. P., Kulichihin, E. P., Mihailova, I. M., & Kerber, M. L. (1996). Rotational and Capillary Viscometry of Melts of Mixes of Traditional and Liquid-Crystalline Thermoplasts. In Abstracts of the 18-th Symposium on Rheology. Karacharovo, 117. [in Russian].
102. Kotomin, S. V., & Kulichihin, B. G. (1999). Flow Limit of Melts of Liquid-Crystal Polyesters and their Mixtures. In Abstracts of the IV Russian Symposium (involving international participants) “Liquid Crystal Polymers”. Moscow, 63. [in Russian].
103. Park Dae Soon, & Kim Seong Hun. (2003). Miscibility Study on Blend of Thermotropic Liquid Crystalline Polymers and Polyester. *J. Appl. Polym. Sci., 87(11),* 1842–1851.
104. Bharadwaj Rishikesh, K., & Boyd Richard, H. (1999). Diffusion of Low-Molecular Penetrant into the Aromatic Polyesters: Modeling with Method of Molecular Dynamics. *Polymer*, *40(15),* 4229–4236.
105. Luscheikin, G. A., Dreval, V. E., & Kulichihin, V. G. (1998). *Vysokomol. Soed. A*-B, *40(9),* 1511–1515.
106. Shumsky, V. F. Getmanchuk, I. P., Rosovitsky, V. F., & Lipatov, Yu. S. (1996). Rheological, Visco-Elastic and Mechanical Properties of Mixes of Polymethylmethacrylate with Liquid-Crystal Copolyester Filled with Wire-like Monocrystals. In Abstracts of the 18-th Symposium on Rheology. Karacharovo, 115. [in Russian].
107. Liu Yongjian, Jin Yi., Dai Linsen, Bu Haishan, & Luise Robert, R. (1999). Crystallization and Melting Behavior of Liquid Crystalline Copolyesters Based on Poly(ethyleneterephthalate). *J. Polym. Sci. A*, *37(3),* 369–377.
108. Abdullaev, Kh. M., Tuichiev, Sh. T., Kurbanaliev, M. K., & Kulichihin, V. G. (1997). *Vysokomol. Soed. A-B, 39(6),* 1067–1070.
109. Li Xin-Gui. (1999). Structure of Liquid Crystalline Copolyesters from two Acetoxybenzoic Acids and Polyethyleneterephthalate. *J. Appl. Polym. Sci., 73(14),* 2921–2925.
110. Li Xin-Gui., & Huang Mei-Rong. (1999). High-Resolution Thermogravimetry of Liquid Crystalline Copoly(p-oxybenzoateethyleneterephthalate-m-oxybenzoate). *J. Appl. Polym. Sci., 73(14),* 2911–2919.
111. Guo Mingming, & Britain William, J. (1998). Structure and Properties of Naphthalene-Containing Polyesters. 4. New Insight into the Relationship of Transesterification and Miscibility. *Macromolecules, 31(21),* 7166–7171.

112. Li Xin-Gui., Huang Mei-Rong, Guan Gui-He., & Sun Tong. (1996). Glass Transition of Thermotropic Polymers Based upon Vanillic Acid, p-hydroxybenzoic Acid, and Poly(ethyleneterephthalate). *J. Appl. Polym. Sci., 59(1),* 1–8.
113. Additives and Modifiers Plast. Compound, *1987–1988(4)*, 10, 14–16, 18, 20, 24, 26, 28, 30, 32, 34, 36, 38–40, 42–44, 46–51.
114. Sikorski, R., & Stepien, A. (1972). Nienasycone Zywice Poliestrowe Zawierajace Cherowiec Cz. *J. Studie Problemowe.–Pr. Nauk. Inst. Technol. Organicz. i Tworzyw. Sztuczn. PWr, (7),* 3–19.
115. Takase, Y., Mitchell, G., & Odajima, A. (1986). Dielectric Behavior of Rigid–Chain Thermotropic Copolyesters. *Polym. Commun, 27(3),* 76–78.
116. Volchek, B. Z., Holmuradov, N. S., Bilibin, A. Yu., & Skorohodov, S. S. (1984). *Vysokomol. Soed. A, 26(1),* 328–333.
117. Bolotnikova, L. S., Bilibin, A. Yu., Evseev, A. K., Panov, Yu. N. Skorohodov, S. S., & Frenkel, S. Ia. (1983). *Vysokomol. Soed. A, 25(10),* 2114–2120.
118. Volchek, B. Z., Holmuradov, N. S., Purkina, A. V., Bilibin, A. Yu., & Skorohodov, S. S. (1984). *Vysokomol. Soed. A, 27(1),* 80–84.
119. Andreeva, L. N., Beliaeva, E. V., Lavrenko, P. N., Okopova, O. P., Tsvetkov, V. N., Bilibin, A. Yu., & Skorohodov, S. S. (1985). *Vysokomol. Soed. A, 27(1),* 74–79.
120. Grigoriev, A. N., Andreeva, L. N., Bilibin, A. Yu., Skorohodov, S. S., & Eskin, V. E. (1984). *Vysokomol. Soed. A, 26(8),* 591–594.
121. Grigoriev, A. N., Andreeva, L. N., Matveeva, G. I., Bilibin, A. Yu., Skorohodov, S. S., & Eskin, V. E. (1985). *Vysokomol. Soed. B, 27(10),* 758–762.
122. Bolotnikova, L. S., Bilibin, A. Yu., Evseev, A. K., Ivanov, Yu. N., Piraner, O. N., Skorohodov, S. S., & Frenkel, S. Ya. (1985). *Vysokomol. Soed. A, 27(5),* 1029–1034.
123. Pashkovsky, E. E. (1986). Abstracts of the Thesis for the Scientific Degree of Candidate of Physical and Mathematical Sciences. Leningrad, 19 [in Russian].
124. Grigoriev, A. N., Andreeva, L. N., Volkov, A. Ya., Smirnova, G. S., Skorohodov, S. S., & Eskin, V. E. (1987). *Vysokomol. Soed. A, 29(6),* 1158–1161.
125. Grigoriev, A. N., Matveeva, G. I., Piraner, O. N., Lukasov, S. V., & Bilibin, A. Yu.,Sidorovich, A. V. (1991). *Vysokomol. Soed. A, 33(6),* 1301–1305.
126. (1981). Liquid Crystal Order in Polymer / Ed. A. Blumshtein. Moscow.
127. Grigoriev, A. N., Matveeva, G. I., Lukasov, S. V., Piraner, O. N., Bilibin, A. Yu., & Sidorovich, A. V. (1990). *Vysokomol. Soed. A-B, 32(5),* 394–396.
128. Bilibin, A. Yu. (1988). *Vysokomol. Soed. B, 31(3),* 163.
129. Kapralova, V. M., Zuev, V. V., Koltsov, A. I., Skorohodov, S. S., & Khachaturov, A. S. (1991). *Vysokomol. Soed. A, 33(8),* 1658–1662.
130. Helfund, E. J. (1971). *Chem. Phys., 54(11),* 4651.
131. Bilibin, A. Yu., Piraner, O. N., Skorohodov, S. S., Volenchik, L. Z., & Kever, E. E. (1990). *Vysokomol. Soed. A, 32(3),* 617–623.
132. Matveeva, G. N. (1986). Abstracts of the Thesis for the Scientific Degree of Candidate of Physical and Mathematical Sciences, 17. [in Russian].
133. Volkov, A. Ya., Grigoriev, A. I., Savenkov, A. D., Lukasov, S. V., Zuev, V. V., Sidorovich, A. V., & Skorohodov, S. S. (1994). *Vysokomol. Soed. B, 36(1),* 156–159.
134. Andreeva, L. N., Bushin, S. V., Matyshin, A. I., Bezrukova, M. A., Tsvetov, V. N., Bilibin, A. Yu., & Skorohodov, S. S. (1990). *Vysokomol. Soed. A, 32(8),* 1754–1759.

135. Stepanova, A. R. (1992). Abstracts of the Thesis for the Scientific Degree of Candidate of Chemical Sciences. Sankt–Petersburg, 24 p. [in Russian].
136. He Xiao-Hua., & Wang Xia-Yu. (2002). Synthesis and Properties of Thermotropic Liquid-Crystalline Block-Copolymers Containing Links of Polyarylate and Thermotropic Liquid-Crystalline Copolyester (HTH-6) *Natur. Sci. J.* Xiangtan Univ, *23(1),* 49–52.
137. Wang Jiu-fen, Zhu-xin, & Huo Hong-Xing. (2003). *J. Funct. Polym, 16(2),* 233–237.
138. Jo Byung-Wook, Chang Jin-Hae., & Jin Jung-2. (1995). Transesterifications in a Polyblend of Poly(butylene terephthalate) and a Liquid Crystalline Polyester *Polym. Eng. and Sci., 35(20),* 1615–1620.
139. Gomez, M. A., Roman, F., Marco, C., Del Pino, J., & Fatou, J. G. (1997). Relaxations in Poly(tetramethylene terephtaloyl-bis-4-oxybenzoate): Effect of Substitution in the Mesogenic Unit and in the Flexible Spacer. *Polymer, 38(21),* 5307–5311.
140. Bilibin, A. Yu., Shepelevsky, A. A., Savinova, T. E., & Skorohodov, S. S. (1982). Terephthaloyl-Bis-*n*-Oxybenzoic Acid or its Dichloranhydride as Monomer for the Synthesis of Thermotropic Liquid-Crystalline Polymers USSR Inventor Certificate 792834. International Patent Catalogue C 07 C 63/06, C 08 K 5/09,
141. Storozhuk, I. P. (1976). *Regularities of the Formation of Poly and Oligoarylenesulfonoxides and Block-Copolymers on their Base.* Thesis for the Scientific Degree of Candidate of Chemical Sciences. Moscow, 195 p. [in Russian].
142. Rigid Polysulfones Hold at 300 F. *Jron. Age,* (1965). *195(15),* 108–109.
143. High-Temperature Thermoplastics. *Chem. Eng. Progr,* (1965). *61(5),* 144.
144. Thermoplastic Polysulfones Strength at High Temperatures. *Chem. Eng. Progr,* (1965). *72(10),* 108–110.
145. Polysulfones (1966). *Brit, Plast, 39(3),* 132–135.
146. Lapshin, V. V. (1967). *Plast. Massy, 1,* 74–78.
147. Gonezy, A. A. (1979). Polysulfon-ein Hochwarmebestandiger, Transparenter Kunststof Kunststoffe, Bild. *69(1),* 12–17.
148. Thornton, E. A. (1968). Polysulfone Thermoplastics for Engineering. *Plast. Eng,*
149. Moiseev, Yu. V., & Zaikov, G. E. (1979). *Chemical Stability of Polymers in Aggressive Media.* Moscow: Himia, 288 p. [in Russian].
150. Thornton, E. A., & Cloxton, H. M. (1968). Polysulfones, Properties and Processing Characteristics *Plastics, 33(364),* 178–191.
151. Huml, J., & Doupovcova, J. (1970). Polysulfon-Nogy Druh Suntetickych Pruskuric. *Plast. Hmoty Akanc, 7(4),* 102–106.
152. Morneau, G. A. (1970). Thermoplastic Polyarylenesulfone that can be Used at 500 °F. *Mod Plast, 47(1),* 150–152, 157.
153. Storozhuk, I. P., & Valetsky, P. M. (1978). *Chemistry and Technology of High-Molecular Compounds, 2,* 127–176.
154. Benson, B. A., Bringer, R. P., & Jogel, H. A. (1967). Polymer 360, a Thermoplastic for Use at 500 °F / Presented at SPE Antes, Detroit, Michigan,
155. Jdem, A. (1967). Phenylene Thermoplastic for Use at 500° F. *SPE Journal,*
156. Besset, H. D., Fazzari, A. M., & Staub, R. B. (1965). Plast. Technol, *11(9),* 50.
157. Jaskot, E. S. (1966). *SPE Journal, 22,* 53.
158. Leslie, V. J. (1974). Properties Et Application Des Polysulfones. *Rev. Gen. Caontch, 51(3),* 159–162.

159. Bringer, R. P., & Morneau, G. A. (1969). Polymer 360, a New Thermoplastic Polysulfone for Use at 500 °F Appl. *Polym. Symp, (11),* 189–208.
160. Andree, U. (1974). Polyarilsulfon Ein Ansergewohnliecher Termoplast Kunststof Kunststoffe, *Bild.* 64, *(11),* S. 684.
161. Giorgi, E. O. (1971). Termoplastico De Engenharia Ideal Para as Condicoes Brasileiras. *Rev. Guim. Ind, 40(470),* 16–18.
162. Korshak, V. V., Storozhuk, I. P., & Mikitaev, A. K. (1976). *Polysulfones–Sulfonyl Containing polymers.* In *Polycondensation Processes and Polymers.* Nalchik, 40–78. [in Russian].
163. Two Tondh Resistant plastic Sthrive in Hot Environments. *Prod. Eng.,* (1969). *40(14),* 112.
164. Polysulfonic Aromatici. *Mater. Plast. Ed Elast,* (1972). *38(12),* 1043–1044.
165. Rose, J. B. (1974). *Polymer, 15(17),* 456–465.
166. Rigby, R. B. (1979). Victrex–Polyestersulfone. *Plast. Panorama Scand, 29(11),* 10–12.
167. Gonozy, A. A. (1979). Polysulfon-ein Hochwarmebeston Dider Transparenter Kunststoff Kunststoffe, Bild. *69(1),* 12–17.
168. Un Nuovo Tecnotermoplastico in Polifenilsulfone Radel. *Mater. Plast Ed Elast,* (1977). *2,* 83–85.
169. Polyestersulfon in Der BASF Palette. Gimmi, Asbest, Kunststoffe. Bild. (1982). *35(3),* 160–161.
170. Bolotina, L. M., & Chebotarev, V. P. (2003). *Plast. Massy, 11,* 3–7.
171. Militskova, A. M., & Artemov, S. V. (1990). *Aromatic Polysulfones, Polyester(Ester) Ketones, Polyphenylenoxides and Polysulphides of NIITEHIM: Review.* Moscow, 1–43. [in Russian].
172. (1999). High-Durable Plastics Kunststoffe, Du Hart Im Nehmen Sind Technica (Susse). Bild. *48(25–26).* 16–22. [in German].
173. Kampf Rudolf. (2006). The Method for Producing Polymers by Means of Condensation in Melt (Polyamides, Polysulfones, Polyarylates etc) Germany Patent Application 102004034708. International Patent Catalogue C 08 P 85/00.
174. Asueva, L. A. (2010). *Aromatic Polyesters Based of Terephthaloyl-Bis-(n-oxybenzoic) Acid.* Thesis for the Scientific Degree of Candidate of Chemical Sciences. Nalchik: KBSU, 129 p. [in Russian].
175. Japan Patent Application 1256524. 1989.
176. Japan Patent Application 1315421. 1995
177. Japan Patent Application 211634. 1990.
178. Japan Patent Application 1256525. 1989.
179. Japan Patent Application 12565269. 1989.
180. Macocinschi Doina, Grigoriu Aurelia, & Filip Daniela. (2002). Aromatic Polyculfones Used for Decreasing of Combustibility. *Eur. Polym. J, 38(5),* 1025–1031.
181. US Patent 6548622. 2003.
182. Synthesis and Characterization Poly(arylenesulfone)s *J. Polym. Sci. A, 2002. 40(4),* 496–510.
183. Germany Patent Application 19926778. 2000.
184. Vologirov, A. K., & Kumysheva, Yu. A. (2003). *Vestnik KBGU. Seria Himicheskih Nauk, (5),* 86. [in Russian].

185. Mackinnon Sean, M., Bender Timothy, P., & Wang Zhi Yuan. (2000). Synthesis and Properties of Polyestersulfones *J. Polym. Sci. A, 38(1),* 9–17.
186. Khasbulatova, Z. S., Asueva, L. A., & Shustov, G. B. (2009). *Polymers on the Basis of Aromatic Oligosulfones* / in *Proceedings of the X International Conference on Chemistry and Physicochemistry of Oligomers*. Volgograd, 100. [in Russian].
187. Ilyin, V. V., & Bilibin, A. Yu. (2002). *Synthesis and Properties of Multiblock-Copolymers Consisting of Flexible and Rigid-Link Blocks* / in Materials of the 3-rd Youth school-Conference on Organic Synthesis. Sankt-Petersburg, 230–231. [in Russian].
188. Germany Patent Application № 19907605. 2000.
189. Reuter Knud, Wollbom Ute., & Pudleiner Heinz. (2000). Transesterification as Novel Method for the Synthesis of Block-Copolymers of Simple Polyester-Sulfone / in Papers of the 38-th Macromolecular IUPAC Symposium. Warsaw, 34.
190. Zhu Shenmin, Xiao Guyu, & Yan Deyue. (2001). Synthesis of Aromatic Graft Copolymers *J. Polym. Sci.* A, *39(17),* 2943–2950.
191. Wu Fangjuan, Song Caisheng, Xie Guangliang, & Liao Guihong. (2007). Synthesis and Properties of Copolymers of 4.4′-Bis-(2-methylphenoxy)Bisphenylsulfone, 1,4-Bisphenoxybenzebe and Terephthaloyl Chloride. *Acta Polym. Sin, 12,* 1192–1195.
192. Ye Su-fang, Yang Xiao-hui, Zheng Zhen, Yao Hong-xi., & Wang Ming-jun. (2006). The Synthesis and Characterization of Novel Aromatic Polysulfones Polyurethane Containing Fluorine, *40(7),* 1239–1243.
193. Ochiai Bundo, Kuwabara Kei., Nagai Daisuke, Miyagawa Toyoharu, & Endo Takeshi. (2006). Synthesis and Properties of Novel Polysulfone Bearing Exomethylene Structure Eur. *Polym. J, 42(8),* 1934–1938.
194. Bolotina L. M., & Chebotarev, V. P. (2007). The Method for Producing the Statistical Copolymers of Polyphenylenesulphidesulfones RF Patent 2311429. *International Patent Catalogue* C 08 G 75/20,
195. Kharaev, A. V., Bazheva, R. Ch., Barokova, E. B., Istepanova, O. L., & Chaika, A. A. (2007). *Fire-Resistant Aromatic Block-Copolymers Based on 1,1-Dichlor-2,2-Bis(n-oxyphenyl)Ethylele* / in Proceedings of the 3-rd Russian Scientific and Practical Conference. Nalchik, 17–21. [in Russian].
196. Saxena Akanksha, Sadhana, R., Rao, V. Lakshmana Ravindran, P. V., & Ninan K. N. (2005). Synthesis and Properties of Poly(ester nitrile sulfone) Copolymers with Pendant Methyl Groups *J. Appl. Polym. Sci., 97,* 1987–1994.
197. Linares, A., & Acosta, J. L. (2004). Structural Characterization of Polymer Blends Based on Polysulfones *J. Appl. Polym. Sci., 92(5),* 3030–3039.
198. Ramazanov, G. A., Shahnazarov, R. Z., & Guliev, A. M. (2005). *Russian. J. Appl. Chem, 78(10),* 1725–1728.
199. Zhao Qiuxia, & Hanson James, E. (2006). Direct Synthesis of Poly(arylmethyl sulfone) Monodendrons. *Synthesis, (3),* 397–399.
200. Cozan, V., & Avram, E. (2003). Liquid-Crystalline Polysulfone Possessing Thermotropic Properties *Eur. Polym. J, 39(1),* 107–114.
201. Dass, N. N. (2000). *Indian J. Phys. A, 74(3),* 295–298.
202. Zhang Qiuyu, Xie Gang, Yan Hongxia, Xiao Jun., & Li Yurhang. (2001). The Effect of Compatibility of Polysulfone and Thermotropic Liquid-Crystalline Polymer *J. North-West. Polytechn. Univ, 19(2),* 173–176.

203. Magagnini, P. L., Paci, M., La Mantia, F. P., Surkova, I. N., & Vasnev, V. A. (1995). Morphology and Rheology of Mixes from Sulfone and Polyester Vectra–A 950 *J. Appl. Polym. Sci., 55(3),* 461–480.
204. Garcia, M., Eguiazabal, J. L., & Nuzabal, J. (2004). Morphology and Mechanical Properties of Polysulfones Modified with Liquid-Crystalline Polymer *J. Macromol. Sci. B, 43(2),* 489–505.
205. RF Patent Application 93003367/04. 1996.
206. Wang Li-jiang, Jian Xi-gao, Liu Yan-jun, & Zheng Guo-dong. (2001). Synthesis and Characterization of Polyarylestersulfoneketone from 1-Methyl-4,5-Bis(chlorbenzoyl)-Cyclohexane and 4-(4-hydroxyphenyl)-2,3-Phthalasin-1-One *J. Funct. Polym, 14(1),* C. 53–56.
207. Lei Wei., & Cai Ming-Zhong. (2004). Synthesis and Properties of Block-Copolymers of Polyesterketoneketone and 4,4{}-Bisphenoxybisphenylsulfone. *J. Appl. Chem, 21(7),* 669–672.
208. Tong Yong-fen, Song Cai-sheng, Wen Hong-li., Chen Lie., & Liu Xiao-ling. (2005). Synthesis and Properties of Copolymers of Arylestersulfones and Esteresterketones Containing Methyl Replacers. *Polym. Mater. Sci. Technol, 21(2),* 162–165.
209. Bowen W. Richard, Doneva Teodora, A., & Yin, H. B. (2000). Membranes Made from Polysulfone Mixed with Polyesteresterketone: Systematic Synthesis and Characterization. Program and Abstr. Tel Aviv, 266.
210. Zinaida S. Khasbulatova, Luisa A. Asueva, Madina A. Nasurova, Arsen M. Karayev, & Gennady B. Shustov. (2006). Polysulfonesterketones on the Oligoester Base, their Thermo-and Chemical Resistance, 99–105.
211. Khasbulatova, Z. S., Asueva, L. A., Nasurova, M. A., Kharaev, A. M., & Temiraev, K. B. (2005). *Simple Oligoesters: Properties and Application.* In Proceedings of the 2-nd Russian Scientific and Practical Conference. Nalchik, 54–57. [in Russian].
212. Khasbulatova, Z. S., Asueva, L. A., Nasurova, M. A., Shustov, G. B., Temiraev, K. B., Kharaeva, R. A., & Asibokova, O. R. (2006). *Synthesis and Properties of Aromatic Oligoesters* / in Materials of International Conference on Organic Chemistry "Organic Chemistry from Butlerov and Belshtein till Nowadays". Sankt-Petersburg, 793–794. [in Russian].
213. Iucкe, A. (1990). Polyarylketone (PAEK) Kunststoffe, Bild. *80(10),* S. 1154–1158, 1063.
214. Khirosi, I. (1983). Polyesterketone Victrex *PEEK, 31(6),* 31–36.
215. Teruo, S. (1982). Properties and Application of Special Plastics. Polyesteresterketone. Koge Dzaire, *30(9),* 32–34.
216. Hay, I. M., Kemmish, D. I., Landford, I. J. and Rae, A. J. (1984). The Structure of Crystalline PEEK. *Polym. Commun, 25(6),* 175–179.
217. Andrew, I. Lovinger, & Davis, D. D. (1984). Single Crystals of Poly (ester-esterketone) (PEEK). *Polym. Commun, 25(6),* 322–324.
218. Wolf, M. (1987). Anwendungstechnische Entwicklungen Bie Polyaromaten Kunststoffe, Bild. *77(6),* S. 613–616.
219. Schlusselindustrien Fur Technische Kunststoffe Plastverarbeiter. 1987. Bild. 38, *(5),* S. 46–47, 50.
220. May, R. (1984). Jn. in Proceedings of the 7-th Anme. Des. Eng. Conf. Kempston, 313–318.

221. Rigby Rhymer, B. (1984). Polyesteretperketone PEEK. *Polymer News, 9,* 325–328.
222. Attwood, T. E., Dawson, P. C., & Freeman, I. L. (1979). Synthesis and Properties of Polyarylesterketones. *Amer. Chem. Soc. Polym. Prepr, 20(1),* 191–194.
223. Kricheldorf, H. R., & Bier, G. (1984). New polymer synthesis 11 Preparation of Aromatic Poly(ester ketone)s from Silylated Bisphenols. *Polymer, 25(8),* 1151–1156
224. (1986). Polyesterketone High. *Polym. Jap., 35(4),* 380
225. (1986). High Heat Resistant Film-Talpa Japan. *Plastics Age, 24(208),* 30.
226. Takao Ia. (1988). Polyestersulfones, Polyesterketones. *Koge Dzaire End Mater, 36(12),* 120–121.
227. Takao Ia. (1990). Polyesterketones. *Koge Dzaire End Mater, 38(3),* 107–116.
228. Khasbulatoba, Z. S., Kharaev, A. M., Miritaev, A. K., et al. (1992). *Plast. Massy, (3),* 3–7.
229. Hergentother, P. M. (1987). Recent Advances in High Temperature Polymers. *Polym. J, 19(1),* 73–83.
230. Method for Producing of the Aromatic Polymer in the Presence of Inert Non-Polar Aromatic Plastificator. US Patent 4110314.
231. Method for Producing Polyesters. Germany Patent Application 2731816.
232. Aromatic Simple Polyesters. GB Patent 1558671.
233. Method for Producing Polyesters. Germany Patent Application 2749645.
234. Producing of Aromatic Simple Polyesters. GB Patent 1569603.
235. Method for Producing Aromatic Polyesters. GB Patent 1563222.
236. Producing of Simple Aromatic Polyesters Containing Microscopic Inclusions of Non-Melting Compounds. US Patent 4331798.
237. Method for Producing Aromatic Polymers. Japan Patent 57–23396.
238. Wear-Resistant, Self-Lubricating Composition. Japan Patent Application 58–109554.
239. Antifriction Composition. Japan Patent Application 58–179262.
240. Thermoplastic Aromatic Polyesterketone. Japan Patent 62–146922.
241. Composition on the Basis of Aromatic Polyarylketones. Japan Patent Application 63–20358.
242. New polyarylketones. US Patent 4731429.
243. Method for Producing Crystalline Aromatic Polyesterketone. US Patent 4757126.
244. Method for Producing High-Molecular Simple Polyesters. Japan Patent Application 63–95230.
245. Aromatic Simple Esters and Method for Producing Same. Japan Patent Application 63–20328.
246. Method for Producing Aromatic Simple Polyesters. Japan Patent Application 63–20328.
247. All-Aromatic Copolyester. Japan Patent Application 63–12360.
248. All-Aromatic copolyesters. Japan Patent Application 63–15820.
249. Colguhoun, H. M. (1984). Synthesis of polyesterketones in Trifluoromethanesulphonic Acid: some Structure-Reactivity Relationships. *Amer. Chem. Soc. Polym. Prepr, 25(2),* 17–18.
250. Polyesterketones. Japan Patent Application 60–144329.
251. Method for Producing Polyesterketones. Japan Patent Application 61–213219.
252. New Polymers and Method for Producing same. Japan Patent Application 62–11726.
253. Method for Producing Simple Polyesterketones. Japan Patent Application 63–75032.

254. Process for Producing Aromatic Polyesterketones. US Patent 4638944.
255. Method for Producing Crystalline Aromatic Simple Polyesterketones. Japan Patent Application 62–7730.
256. Method for Producing Thermoplastic Aromatic Simple Polyesters. Japan Patent Application 62–148524.
257. Method for Producing Thermoplastic Polyesterketones. Japan Patent Application 62–148323.
258. Thermoplastic Aromatic Simple Polyesterketones and Method Producing Same. Japan Patent Application 62–151421.
259. Method for Producing Polyarylesterketones Using Catalyst on the Basis of Sodium Carbonate and Salt of Organic Acid. US Patent 4748227.
260. Thermostable Polyarylesterketones. Germany Patent Application 37008101.
261. Simple Aromatic Polyesterketones. Japan Patent Application 63–120731.
262. Impact Strength Polyarylesterketones. Japan Patent Application 63–120730.
263. Method for Producing Simple Polyarylesterketones in the Presence of Salts of Lanthanides, *Alkali and Alkali-Earth Metals.* US Patent 4774311.
264. Jovu, M., & Marinecsu, G. Rolicetoeteri. (1981). Produce de Policondensaze Ale 4,4–Dihidroxibenzofenonei Cu Compusi Bisclorometilate Aromatici *Rev. Chim, 32(12),* 1151–1158.
265. Sankaran, V., & Marvel, C. S. (1979). Polyaromatic Ester-Ketone-Sulfones Containing 1,3-Butadiene Units *J. Polymer Sci.: Polymer Chem. Ed, 17(12),* 3943–3957.
266. Method for Producing Aromatic Polyesterketones. Japan Patent Application 60–101119.
267. Method for Producing Aromatic Polyesterketones and Polythioesterketones. US Patent 4661581.
268. Method for Producing Aromatic Polyesterketones. Germany Patent Application 3416446.
269. Uncrosslinked-Linked Thermoplastic Reprocessible Polyesterketone and Method for its Production. Germany Patent Application 3416445 A.
270. Litter, M. J., & Marvel, C. S. (1986). Polyaromatic Esterketones and Polyaromatic Ester-Ketone Sulfonamides from 4-Phenoxy-Benzoyl Ester. *J. Polym. Sci.: Polym. Chem. Ed, 23(8),* 2205.
271. Method for Producing Aromatic Simple Poly(thio)Esterketone. Japan Patent Application 61–221228.
272. Method for Producing Aromatic Simple Poly(thio)Esterketone. Japan Patent Application 61–221229.
273. Method for Producing Polyarylketone Involving Treatment with the Diluents. US Patent 4665151.
274. Copolyesterketones. US Patent 4704448.
275. Method for Producing Aromatic Poly(thio)Esterketones. Japan Patent Application 62–146923.
276. Method for Producing Simple Aromatic Polythioesterketones. Japan Patent Application 62–119230.
277. Production of Aromatic Polythioesterketones. Japan Patent Application 62–241922.
278. Method for Producing Polyarylenesterketones. US Patent 4698393.
279. Method for Producing Aromatic Polymers. US Patent 4721771.

280. Method for Producing Aromatic Simple Poly(thio)Esterketones. Japan Patent Application 63–317.
281. Method for Producing Aromatic Simple Poly(thio)Esterketones. Japan Patent Application 63–316.
282. Method for Producing Polyarylesterketones. US Patent 471611.
283. Gileva, N. G., Solotuchin, M. G., & Salaskin, S. N. (1988). Synthese Von Aromatischen Polyketonen Durch Fallungspolukondensation. *Acta Polym*, Bild. 39, *(8)*, S. 452–455.
284. Lee, I., & Marvel, S. (1983). Polyaromatic Esterketones from o,o-Disubstituted Diphenyl Esters *J. Polym. Sci.: Polym. Chem. Ed*, *21(8)*, 2189–2195.
285. Method for Producing Polyarylenesterketones by Means of Electrophylic Polycondensation. Germany Patent Application 3906178.
286. Colgupoum, H. M., & Lewic, D. F. (1988). Aromatic Polyesterketones Via Superacid Catalysis / in "Spec. Polym. 88": Abstracts of the 3-rd International Conference on. *New Polymeric Materials*. Guildford, 39.
287. Colgupoum, H. M., & Lewic, D. F. (1988). Synthesis of Aromatic Polyester-Ketones in Triflouromethanesulphonic Acid. *Polym*, *29(10)*, 1902.
288. Durvasula, V. R., Stuber, F. A., & Bhattacharyee, D. (1988). Synthesis of Polyphenyleneester and Thioester Ketones. *J. Polym. Sci. A*, *27(2)*, 661–669.
289. Method for Producing High-Molecular Polyarylenesulphideketone. US Patent 47182122.
290. Ogawa, T., & Marvel, C. S. (1985). Polyaromatic Esterketones and Ester-Ketone-Sulfones having Various Hydrophilic Groups. *J. Polym. Sci.: Polym. Chem.* Ed, *23(4)*, 1231–1241.
291. Percec, V., & Nava, H. (1988). Synthesis of Aromatic Polyesters by Scholl Reaction l, Poly(1,1-Dinaphthyl Ester Phenyl Ketones). *J. Polym. Sci. A*, *26(3)*, 783–805.
292. Mitsuree, U., & Nasaki, S. (1987). Synthesis of Aromatic Poly(ester ketones) Macromolecules, *20(11)*, 2675–2677.
293. Method for Producing Simple Polyesterketones. Japan Patent Application 61–247731.
294. Aromatic Polyesterketone and its Production. Japan Patent Application 61–143438.
295. Crystalline Polymers with Aromatic Ketone, Simple Ether and Thioether Linkages within the Main Chain and Method for Producing Same. Japan Patent Application 61–141730.
296. Producing of Crystalline Aromatic Polysulphidesterketones. Japan Patent Application 62–529.
297. Producing of Crystalline Aromatic Polyketone with Simple Ether and Sulphide Linkages. Japan Patent Application 62–530.
298. Patel, H. G., Patel, R. M., & Patel, S. R. (1987). Polyketothioesters from 4,4-Dichloroacetyldiphenylester and their Characterization. *J. Macromol. Sci. A*, *24(7)*, 835–340.
299. Method for Producing Aromatic Polyesterketones. Japan Patent Application 62–220530.
300. Method for Producing Aromatic Polyesterketones. Japan Patent Application 62–91530.

301. Aromatic Copolyketones and Method for Producing Same. Japan Patent Application 63–10627.
302. Crystalline Aromatic Polyesterketones and Method for Producing same. Japan Patent Application 61–91165.
303. Aromatic Polyesterthioesterketones and Method for Producing same. Japan Patent Application 61–283622.
304. Producing of Polyarylenoxides using Carbonates of Alkali-Earth Metals, Salts of Organic Acids and, in Some Cases, salts of Copper as catalysts. US Patent 4774314.
305. Method for Producing Polyarylenesterketones US Patent 4767837.
306. Heat-Resistant Polymer and Method for Producing same. Japan Patent Application 62–253618.
307. Heat-Resistant Polymer and Method for Producing Same. Japan Patent Application 62–253619.
308. Aromatic Polyesterketones. US Patent 4703102.
309. Producing of Aromatic Polymers. GB Patent 1569602.
310. New Polymers and Method for their Production. Japan Patent Application 61–28523.
311. Aromatic Simple Polyesterketones with Blocked end Groups and Method for Producing Same Japan Patent Application 61–285221.
312. Aromatic Simple polyesterketones and Method for their Production. Japan Patent Application 61–176627.
313. Method for Producing Crystalline Aromatic simple Polyesterketones. Japan Patent Application 62–7729.
314. Method for Producing Fuse Aromatic Polyesters. US Patent 4742149.
315. Films from Aromatic Polyesterketones Germany. Patent Application 3836169.
316. Method for Producing Polyarylenestersulfones and Polyarylenesterketones. Germany Patent Application 3836582.
317. Method for Producing Polyarylenesterketones. Germany Patent Application 3901072.
318. Polyarylesterketones. US Patent 4687833.
319. Method for Producing Oligomer Aromatic Simple Ethers. Poland Patent 117224.
320. Polymers Containing Aromatic Groupings. GB Patent 1541568.
321. Corfield, G. C., & Wheatley, G. W. (1988). The Synthesis and Properties of Blok Copolymers of Polyesteresterketone and Polydimethylsiloxane. In "Spec. Polym. 88": Abstracts of the 3-rd International Conference on New Polymeric Materials. Cambridge, 68.
322. Method for Producing Polyarylesterketones. Germany Patent Application 3700808.
323. Block-Copolymers Containing Polyarylesterketones and Methods for their Production. US Patent 4774296.
324. Poly(arylesterketones) of Improved Chain. US Patent 4767838.
325. New Block-Copolymer Polyarylesterketone-Polyesters. US Patent 4668744.
326. Simple Polyarylesterketone Block-Copolymers. US Patent 4861915.
327. Producing of Polyarylenesterketones by Means of Consecutive Oligomerization and polycondensation in Separate Reaction Zones. US Patent 4843131.
328. Khasbulatova, Z. S., Kharaev, A. M., Mikitaev, A. K., et al. (1990). *Plast Massy, (11),* 14–17.

329. Khasbulatova, Z. S. (1989). *Diversity of Methods for Synthesizing Polyesterketones* / in Abstracts of the II Regional Conference "Chemists of the Northern Caucasus–to National Economy". Grozny, 267. [in Russian].
330. Reimer Wolfgang. (1999). Polyarylesterketone (PAEK) Kunststoffe, Bild. 89, *(10),* S. 150, 152, 154.
331. Takeuchi Hasashi, Kakimoto Masa-Aki., & Imai Yoshio. (2002). Novel Method for Synthesizing Aromatic Polyketones from Bis(arylsilanes) and Chlorides of Aromatic Bicarbonic Acids *J. Polym. Sci. A, 40(16),* 2729–2735.
332. Process for Producing Polyketones. US Patent 6538098. International Patent Catalogue C 08 П 6/00, 2003.
333. Maeyata Katsuya, Tagata Yoshimasa, Nishimori Hiroki, Yamazaki Megumi, Maruyama Satoshi, & Yonezawa Noriyuki. (2004). Producing of aromatic polyketones on the Basis of 2,2{}-Diaryloxybisphenyls and Derivatives of Arylenecarbonic Acids Accompanied with Polymerization with Friedel-Krafts Acylation React. *And Funct. Polym, 61(1),* 71–79.
334. Daniels, J. A., & Stephenson, J. R. (1995). Producing of Aromatic Polyketones. GB Patent Application 2287031. International Patent Catalogue C 08 G 67/00,
335. Gibeon Harry, W., & Pandya Ashish. (1994). Method for Producing Aromatic Polyketones. US Patent 5344914. International Patent Catalogus C 08 G 69/10,
336. Zolotukhin, M. G., Baltacalleja, F. J., Rueda, D. R., & Palacios, J. M. (1997). Aromatic Polymers Produced by Precipitate Condensation *Acta Polym, 48(7),* 269–273.
337. Zhang Shanjy, Zheng Yubin, Ke Yangchuan, & Wu Zhongwen. (1996). Synthesis of Aromatic Polyesterketones by Means of Low-Temperature Polycondensation Acta Sci. *Nature. Univ. Jibimensis, 1,* 85–88.
338. Hachya Hiroshi, Fukawa Isaburo, Tanabe Tuneaki, Hematsu Nobuyuki, & Takeda Kunihiko. (1999). Chemical Structure and Physical Properties of Simple Polyesterketone Produced from 4,4′-Dichlorbenzophenone and Sodium Carbonate Trans. *Jap. Soc. Mech. Eng.* A, *65(632),* 71–77.
339. Yang Jinlian, & Gibson Harry W. (1997). Synthesis of Polyketones Involving Nucleophylic Replacement Through Carb-Anions Obtained From Bis(α-aminonitriles) Macromolecules, *30(19),* 64–73.
340. Yang Jinlian, Tyberg Christy S., & Gibson Harry W. (1999). Synthesis of Polyketone Containing Nucleophylic Replacers Through Carb-Anions Obtained from Bis(α-aminonitriles). Aromatic polyesterketones Macromolecules, *32(25),* 8259–8268.
341. Yonezawa Noriyuki, Ikezaki Tomohide, Nakamura Niroyuki, & Maeyama Katsuya. (2000). Successful Synthesis of all-Aromatic Polyketons by means of Polymerization with Aromatic Combination in the Presence of Nickel *Macromolecules, 33(22),* 8125–8129.
342. (2005). Aromatic Polyesterketones US Patent 6909015. International Patent Catalogue C 07 C 65/00.
343. Toriida Masahiro, Kuroki Takashi, Abe Takaharu, Hasegawa Akira, Takamatsu Kuniyuki, Taniguchi Yoshiteru, Hara Isao, Fujiyoshi Setsuko, Nobori Tadahito, & Tamai Shoji. (2004). Patent Applicaiton 1464662. International Patent Catalogue C 08 G 65/40.

344. Richter Alexander, Schiemann Vera, Gunzel Berna, Jilg Boris, & Uhlich Wilfried. (2007). Verfahren Zur Herstellung Von Polyarylenesterketon Germany Patent Application 102006022442. International Patent Catalogue C 08 G 65/40.
345. Chen Liang, Yu Youhai, Mao Huaping, Lu Xiaofeng, Yao Lei., & Zhang Wanjin. (2005). Synthesis of a new Electroactive Poly(aryl ester ketone) Polymer, *46(8)*, 2825–2829.
346. Maikhailin Yu. A. (2007). *Polymer. Mater: Articles, Equip. Technol, 5,* 6–15.
347. Sheng Shouri, Kang Yigiang, Huang Zhenzhong, Chen Guohua, & Song Caisheng. (2004). Synthesis of Soluble Polychlorreplaced Polyarylesterketones Acta Polym. *Sin, 5,* 773–775.
348. Kharaev, A. M., Mikitaev, A. K., & Bazheva, R. Ch. (2007). *Halogen-Containing Polyarylenesterketones* / in Proceedings of the 3-rd Russina Scientific and Practical Conference "Novel Polymeric Composite Materials". Nalchik, 187–190. [in Russian].
349. Liu Baijun, Hu Wei., Chen Chunhai, Jiang Zhenhua, Zhang Wanjin, Wu Zhongwen, & Matsumoto Toshihik. (2004). Soluble Aromatic Poly(ester ketone)s with a Pendant 3,5-Ditrifluoromethylphenyl Group *Polymer, 45(10),* 3241–3247.
350. Gileva, N. G., Zolotukhin, N. G., Sedova, E. A., Kraikin, V. A., & Salazkin, S. N. (2000). *Synthesis of Polyarylenephthalidesterketones* / in Abstracts of the 2-nd Russian Kargin Symposium Chemistry and Physics of Polymers in the Beginning of the 21 Century. *Chernogolovka, Part 1. P. 1/83.* [in Russian].
351. Wang Dekun, Wei Peng, & Wu Zhe. (2000). Synthesis of Soluble Polyketones and Polyarylenevinylens–new Reaction of Polymerization *Macromolecules, 33(18),* 6896–6898.
352. Wang Zhonggang, Chen Tianlu, & Xu Jiping. (1995). Synthesis and Characteristics of Card Polyarylesterketones with Various Alkyl Replacers. *Acta Polym. Sin, 4,* 494–498.
353. Salazkin, S. N., Donetsky, K. I., Gorshkov, G. V., Shaposhnikova, V. V., Genin, Ya. V., & Genina, M. M. (1997). *Vysokomol. Soed. A-B, 39,* C. 1431–1437.
354. Salazkin, S. N., Donetsky, K. I., Gorshkov, G. V., & Shaposhnikova, V. V. (1996). *Doklady RAN, 348(1),* C. 66–68.
355. Donetsky, K. I. (2000). Abstracts of the Thesis for the Scientific Degree of Candidate of Chemical Sciences. Moscow, 24 p. [in Russian].
356. Khalaf Ali, A., Aly Kamal, L., & Mohammed Ismail, A. (2002). New Method for Synthesizing Polymers *J. Macromol. Sci. A, 39(4),* 333–350.
357. Khalaf Ali, A., & Alkskas, I. A. (2003). Method for Synthesizing Polymers. *Eur. Polym. J, 39(6),* 1273–1279.
358. Aly Kamal, L. (2004). Synthesis of Polymers *J. Appl. Polym. Sci., 94(4),* .1440–1448.
359. Chu, F. K., & Hawker, C. J. (1993). Different Syntheses of Isomeric Hyperbranched Polyesterketones. *Polym. Bull, 30(3),* 265–272.
360. Yonezawa Noriyuki, Nakamura Hiroyuki, & Maeyama Katsuya. (2002). Synthesis of all-Aromatic Polyketones having Controllable Isomeric Composition and Containing Links of 2-Trifluorometylbisphenylene and 2,2{}-Dimetoxybisphenylene. *React. And Funct. Polym, 52(1),* 19–30.

361. Zhang Shaoyin, Jian Xigao, Xiao Shude, Wang Huiming, & Zhang Jie. (2002). Synthesis and Properties of Polyarylketone Containing Bisphthalasinone and Methylene Groupings. *Acta Polym. Sin, 6,* 842–845.
362. Chen Lianzhou, Jian Xigao, Gao Xia., & Zhang Shouhai. (1999). Synthesis and Properties of Polyesterketones Containing Links of Chlorphenylphthalasion. *Chin. J. Appl. Chem, 16(3),* 106–108.
363. Gao Ye., & Jian Xi-gao. (2001). Synthesis and Crharacterisation of Polyearyesterketones Contnaining 1,4-Naphthaline Linkages *J. Dalian Univ. Technol, 41(1),* 56–58.
364. Wang Mingjing, Liu Cheng, Liu Zhiyong, Dong Liming, & Jian Xigao. (2007). Synthesis and Properties of Polyarylnithilesterketoneketones Containing Phthalasinon *Acta Polym. Sin, 9,* 833–837.
365. Zhang Yun-He., Wang Dong, Niu Ya-Ming, Wang Gui-Bin., & Jiang Zhen-Hua. (2005). Synthesis and Properties of Fluor-Containing Polyarylesterketones with Links of 1,4-Naphthylene. *Chem. J. Chin. Univ, 26(7),* 1378–1380.
366. Kim Woo-Sik., & Kim Sang-Youl. (1997). Synthesis and Properties of Polyesters Containing Naphthalenetetracarboxylic Imide. *Macromol. Symp, (118),* 99–102.
367. Cao Hui., Ben Teng, Wang Xing, Liu Na., LiuXin-Cai., Zhao Xiao-Gang, Zhang Wan-Jin., & Wei Yen. (2004). Synthesis and Properties of Chiral Polyarylesterketones Containing Links of 1,1{}-Bis-2-Naphtyl. *Chem. J. Chin. Univ, 25(10),* 1972–1974.
368. Wang Feng, Chen Tianlu, Xu Jiping, Lui Tianxi, Jiang Hongyan, Qi Yinhua, Liu Shengzhou, & Li Xinyu. (2006). Synthesis and Characterization of Poly(arylene ester ketone) (co)Polymers Containing Sulfonate Groups. *Polymer, 47(11),* 4148–4153.
369. Cheng Cai-Xia., Liu-Ling, & Song Cai-Sheng. (2002). Synthesis and Properties of Aromatic Polyesterketoneketone Containing Carboxylic Group within the Lateral. *Chain J. Jiangxi Norm. Univ. Natur. Sci. 26(1),* 60–63.
370. 2,3,4,5,6-Pentafluorobenzoylbisphenylene Ethers and Fluor-Containing Polymers of Arylesterketones US Patent 6172181. International Patetn Catalogue C 08 Π 73/24, 2001.
371. Ash, C. E. (1995). Process for Producing Stabilized Polyketones US Patent 5432220. International Patent Catalogue C 08 F 6/00.
372. Jiang Zhen-yu., Huang Hai-Rong, & Chen Jian-Ding. (2007). Synthesis and Properties of Polyarylesterketone and Polyarylestersulfone Containing Links of Hexafluoroizopropylydene. *J. E. China Univ. Sci.* and *Technol. Nat. Sci. 33(3),* 345–349.
373. Xu Yongshen, Gao Weiguo, Li Hongbing, & Guo Jintang. (2005). Synthesis and Properties of Aromatic Polyketones Based on CO and Stirol or *n*-ethylstirol *J. Chem. Ind. Eng.* (China)., *56(5),* 861–864.
374. Rao, V. L., Sabeena, P. U., Saxena Akanksha, Gopalakrishnan, C., Krishnan, K., Ravindran, P. V., & Ninan, K. N. (2004). Synthesis and Properties of Poly(aryl ester ester ketone) Copolymers with Pendant Methyl Groups. *Eur. Polym. J, 40(11),* 2645–2651.
375. Tong Yong-Fen., Song Cai-Sheng, Chen Lie., Wen Hong-Li., & Liu Xiao-Ling. (2004). Synthesis and Properties of Methyl-Replaced Polyarylesterketone. *Chin. J. Appl. Chem, 21(10),* 993–996.
376. Koumykov, R. M., Vologirov, A. K., Ittiev, A. B., & Rusanov, A. L. (2005). *Simple Aromatic Polyesters and Polyesterketones Based on Dinitro-Derivatives of Clroral*

/ in "Novel Polymeric Composite Materials": Proceedings of the 2-nd Russian Research-Practical Conference. Nalchik, 225–228. [in Russian].
377. Koumykov, R. M., Bulycheva, E. G., Ittiev, A. B., Mikitaev, A. K., & Rusanov, A. L. (2008). *Plast. Massy, 3,* 22–24.
378. Polyester Ketone and Method of Producing the Same US Patent 7217780. International Patent Catalogue C 08 G 14/04. 2006.
379. Li Jianying, Yu Yikai, Cai Mingzhong, & Song Caisheng. (2006). Synthesis and Properties of Simple Polyesterketonesterketone Containing Lateral Cyanogroups. *Petrochem. Technol, 35(12),* 1179–1183.
380. Liu Dan., & Wang Zhonggang. (2008). Novel Polyaryletherketones Bearing Pendant Carboxyl Groups and their Rare Earth Complexes. Part I. Synthesis and Characterization. *Polymer, 49(23),* 4960–4967.
381. Jeon In-Yup., Tan Loon-Seng, & Baek Jong-Beom. (2007). Synthesis of Linear and Hyperbranched Poly(esterketone)s Containing Flexible Oxyethylene Spacers. *Polym. Sci. A, 45(22),* 5112–5122.
382. Maeyama Katsuya, Sekimura Satoshi, Takano Masaomi, & Yonezawa Noriyuki. (2004). Synthesis of Copolymers of Aromatic Polyketones React. And Funct. Polym, *58(2),* 111–115.
383. Li, Wei., Cai Ming-Zhong, & Song Cai-Sheng. (2002). Synthesis of Ternary Copolymers from 4,4′-Bisphenoxybisphenylsulfone, 4,4′-Bisphenoxybenzophenone and Terephthaloylchloride. *Chin. J. Appl. Chem, 19(7),* 653–656.
384. Gao Yan., Dai Ying, Jian Xigao, Peng Shiming, Xue Junmin, & Liu Shengjun. (2000). Synthesis and Characterization of Copolyesterketones Produced from Hexaphenyl-Replaced Bisphenylbisphenol and Hydroquinone. *Acta Polym. Sin, 3,* 271–274.
385. Sharapov, D. S. (2006). Abstracts of the Thesis for the Scientific Degree of Candidate of Chemical Sciences. Moscow, 25 p. [in Russian].
386. Kharaeva, R. A., & Ashibokova, O. R. (2005). *Synthesis and Some Properties of Copolyesterketones* / in Proceedings of Young Scientists. Nalchik, KBSU, 138–141. [in Russian].
387. Method for Preparing Polyester Copolymers with Polycarbonates and Polyarylates. US Patent 6815483. International Patent Catalogue C 08 L 67/00. 2004.
388. Liu Xiao-Ling, Xu Hai-Yun., & Cai Ming-Zhong. (2001). Synthesis and Properties of Statistical Copolymers of Polyesterketoneketone and Polyesterketoneesterketoneketone Containing Naphthalene Cycle Wothon the Main Chain *J. Jiangxi Norm. Univ. Natur. Sci.* Ed, *25(4),* 292–294.
389. Synthesis and Properties of Poly(aryl ester ketone) Copolymers Containing 1,4-Naphthalene Moieties. (2004). *J. Macromol. Sci. A, 41(10),* 1095–1103.
390. Yu Yikai, Xiao Fen., & Cai Mingzhong. (2007). Synthesis and Properties of Poly(arylesterketone ketone)/poly(aryl ester ester ketone ketone) Copolymers with Pendant Cyano Groups. *J. Appl. Polym. Sci., 104(6),* 3601–3606.
391. Tong Yong-fen, Song Cai-sheng, Chen Lie., Wen Hong-li., & Liu Xiao-ling. (2005). Synthesis and Properties of Copolymers of Polyarylesterketone Containing Lateral Methyl Groupings. *Polym Mater. Sci. Technol. Eng, 21(4),* 70–72, 76.
392. Gao Yan., Robertson Gilles, P., Guiver Michael, D., Mikhailenko Serguei D., Li Xiang, & Kaliaguine Serge. (2004). Synthesis of Copolymers of Polyaryleneesteresterketoneketones Containing Links of Naphthalene Sulfonic Acid within the Lat-

eral Links, and their user at Manufacturing Proton-Exchange Membranes. *Macromolecules, 37(18),* 6748–6754.
393. Mohwald Helmut, Fischer Andreas, Frambach Klaus, Hennig Ingolf, & Thate Sven. (2004). Verfahren zur Herstellung Eines Zum Protonenaustausch Befahigter Polymersystems Auf Der Basis Von Polyarylesterketonen Germany Patent Application 10309135. International Patent Catalogue C 08 G 8/28,
394. Shaposhnikova, V. V., Sharapov, D. S., Kaibova, I. A., Gorlov, V. V., Salazkin, S. N., Dubrovina, L. V., Bragina, T. P., Kazantseva, V. V., Bychko, K. A., Askadsky, A. A., Tkachenko, A. S., Nikiforova, G. G., Petrovskii, P. V., & Peregudov, A. S. (2007). *Vysokomol. Soed, 49(10),* 1757–1765.
395. Shaposhnikova, V. V., Salazkin, S. N., Matedova, I. A., & Petrovskii, P. V. Polyarylenetherketones. Investigation of Approaches to Synthesis of Amorphous Blockpolymers in Abstracts of the 4-th International Symposium Molecular Order and Mobility in Polymer Systems. – St. Petersburg. – . 121.
396. Bedanokov, A. Yu. (1999). Abstracts of the Thesis for the Scientific Degree of Candidate of Chemical Sciences. Nalchik, 19 p. [in Russian].
397. Yang Yan-Hua., Dai Xiao-Hui., Zhou Bing, Ma Rong-Tang, & Jiang Zhen-Huang. (2005). Synthesis and Characterization of Block Copolymers Containing Poly(aryl ester ketone) and Liquid Crystalline Polyester Segments. *Chem. J. Chin. Univ, 26(3),* 589–591.
398. Zhang Yun-He., Liu Qin-Hua., Niu Ya-Ming, Zhang Shu-Ling, Wang Dong, & Jiang Zhen-Hua. (2005). Properties and Crystallization Kinetics of Poly(ester ester ketone)-Co-Poly(ester ester ketone ketone) Block Copolymers. *J. Appl. Polym. Sci., 97(4),* 1652–1658.
399. Polyarylenesterketone Phosphine Oxide Compositions Incorporation Cycloaliphatic Units for Use as Polymeric Binders in Thermal Control Coatings and Method for Synthesizing Same US Patent 7208551. International Patent Catalogue C 08 L 45/00 216.
400. Keshtov, M. L., Rusanov, A. L., Keshtova, S. V., Pterovskii, P. V., & Sarkisyan, G. B. (2001). *Vysokomol. Soed. A, 43(12),* 2059–2070.
401. Keshtov, M. L., Rusanov, A. L., Keshtova, S. V., Schegolihin, A. N., & Petrovskii, P. V. (2001). *Vysokomol. Soed. A, 43(12),* 2071–2080.
402. Brandukova, Natalya, E., & Vygodskii Yakov, S. (1995). Novel Poly-α-Diketones and Copolymers on their Base *J. Macromol. Sci. A, 32,* 941–950.
403. Yandrasits, M. A., Zhang, A. Q., Bruno, K., Yoon, Y., Sridhar, K., Chuang, Y. W., Harris, F. W., & Cheng, S. Z. D. (1994). Liquid-Crystal Polyenamineketones Produced Via Hydrogen Bonds. *Polym. Int, 33(1),* 71–77.
404. Mi Yongli, Zheng Sixun, Chan Chi-ming, & Guo Qipeng. (1998). Mixes of Phenolphthalenin with Thermotropic Liquid-Crystal Copolyester. *J. Appl. Polym. Sci., 69(10),* 1923–1931.
405. Arjunan Palanisamy. (1995). Production of Polyesters from Polyketones US Patent 5466780. International Patent Catalogue C 08 F 8/06, C 08 K 5/06.
406. Arjunan Palanisamy. (1996). Process of Transformation of Polyketones into the Complex Polyesters US Patent 55506312. International Patent Catalogue C 08 F 20/00.

407. Matyushov, V. F., & Golovan' S. V. (2003). Method for Producing Non-Saturated Oligoarylesterketones RF Patent 2201942. International Patent Catalogue C 08 G 61/12.
408. Matyushov, V. F., & Golovan, S. V. (2003). Method for Producing Non-Saturated Oligoarylesterketones RF Patent Application 2001109440/04. International Patent Catalogue C 08 G 61/12.
409. Matyushov, V. F., Golovan, S. V., & Malisheva, T. L. (2000). Method for Producing Oligoarylesterketones with End Amino-Groups Ukraine Patent 28015. International Patent Catalogue C 08 G 8/02.
410. Zhaobin Qiu., Zhishen Mo., & Hongfang Zhang. (2000). Synthesis and Crystalline Structure of Oligomer of Arylesterketone. *Chem Res, 11,* 5–7.
411. Guo Qingzhong, & Chen Tianlu. (2004). Synthesis of Macrosyclic Oligomers of Aryleneketones Containing Phthaloyl Links by Measn of Friedel-Crafts Acylation Reaction Chem. Lett, *33(4),* 414–415.
412. Wang Hong Hua., Ding Jin., & Chen Tian Lu. (2004). Cyclic Oligomers of Phenolphthalein Polyarylene Ester Sulfone (ketone): Preparation Through Cyclo-Depolymerization of Corresponding Polymers. *Chin. Chem. Lett, 15(11),* 1377–1379.
413. Kharaev, A. M., Basheva, R. Ch., Istepanova, O. L., Istepanov, M. I., & Kharaeva, R. A. (2006). Aromatic Oligoesterketones for Polycondensation RF Patent 2327680. International Paten Catalogue C 07 C 43/02,
414. Bedanokov A. Yu., Shaov A. Kh., Kharaev A. M., & Dorofeev V. T. (2000). *Plast. Massy, 4,* 42.
415. Bedanokov Azamat, U., Shaov Abubekir, Ch., Charaev Arsen, M., & Mashukov Nurali, I. (1997). *Sythesis and Some Properties of Oligo-and Polyesterketones Based on Bisphenylpropane* / in Proceedings of International Symposium New Approaches in Polymeric Syntheses and Macromolecular Formation. Sankt-Petersburg, 13–17. [in Russian].
416. Kharaev, A. M., Bazheva, R. Ch., Kazancheva, F. K., Kharaeva, R. A., Bahov, R. T., Sablirova, E. R., & Chaika, A. A. (2005). *Aromatic Polyesterketones and Polyesteresterketones as Perspective Thermostable Constructional Materials* / in Proceedings of the 2-nd Russian Research-Practical Conference. Nalchik, 68–72. [in Russian].
417. Kharaev, A. M., Bazheva, R. Ch., Kharaeva, R. A., Beslaneeva, Z. L., Pampuha, E. V., & Barokova, E. B. (2005). *Producing of Polyesterketones and Polyesteresterketones on the Basis of Bisphenols of Various Composition* / in Proceedings of the 2-nd Russian Research-Practical Conference. Nalchik, 44–47. [in Russian].
418. Bazheva, R. Ch., Kharaev, A. M., Olhovaia, G. G., Barokova, E. B., & Chaika, A. A. (2006). Polyester-Polyesterketone Block-Copolymers / in Abstracts of the International Conference on Organic Chemistry "Organic Chemistry from Butlerov and Belshtein Till Nowadays". Sankt-Petersburg, 716. [in Russian].
419. Aromatic Polymers. GB Patent 1563223.
420. Polysulfoneesterketones Germany Patent Application 3742445.
421. Germany Patent Application 3742264.
422. Aromatic Polymers. *Macromolecules,* (1984). *17(1),* 10–14.
423. Khasbulatova, Z. S., Kharaev, A. M., & Mikitaev, A. K. (2009). Khim. *Prom. Segodnya, 10,* 29–31.

424. Wen Hong-Li., Song Cai-Sheng, Tong Yong-Fen., Chen Lie., & Liu Xiao-Ling. (2005). Synthesis and Properties of Poly(aryl ester sulfone ester ketone ketone) (PESEKK) *J. Appl. Polym. Sci., 96(2),* 489–493.
425. Li Wei., & Cai Ming-Zhong Ying. (2004). *Chin. J. Appl. Chem, 21(7),* 669–672.
426. Sheng Shou-Ri., Luo Qiu-Yan.,Yi-Huo., Luo Zhuo, Liu Xiao-Ling, & Song Cai-Sheng. (2008). Synthesis and Properties of Novel Organosoluble Aromatic Poly(ester ketone)s Containing Pendant Methyl Groups and Sulfone Linkages *J. Appl. Polym. Sci., 107(1),* 683–687.
427. Tong Yong-fen, Song Cai-sheng, Wen Hong-li., Chen Lie., Liu Xiao-Ling. (2005). Synthesis and Properties of Copolymers Containing Methyl Replacers *Polym. Mater. Sci. Technol, 21(2),* 162–165.
428. Sheng Shou-ri., Luo-Qiu-yan, Huo Yi., Liu Xiao-ling, Pei Xue-liang, & Song Cai-sheng. (2006). Synthesis and Properties of Soluble Methyl-Replaced Polyarylesterketonestersulfonesterketones. *Polym. Mater. Sci. Technol, 22(3),* 85–87, 92.
429. Xie Guang-Liang, Liao Gui-Hong, Wu Fang-Juan, & Song Cai-Sheng. (2008). Synthesis and Adsorption Properties of Poly(arylestersulfonesterketone)Ketone with Lateral Carboxylic Groups. *Chin. J. Appl. Chem, 25(3),* 295–299.
430. Charaev, A. M., Khasbulatova, Z. S., Basheva, R. Ch., Kharaeva, R. A., Begieva, M. B., Istepanova, O. L., & Istepanov, M. I. (2007). *Izv. Vuzov. Sev.-Kav. Reg. Estestv. Nauki, 3,* 50–52.
431. Chen Lie., Song Cai-Sheng, Wen Hong-Li., Tong Yong-Fen., & Liu Xiao-Ling. (2004). Synthesis of Statistical Polyestersulfonesterketoneketones Containing Bis(o-methyl) Groups. *Chin. J. Appl. Chem, 21(12),* 1245–1248.
432. Arthanareeswaran, G., Mohan, D., & Raajenthiren, M. (2007). Preparation and Performance of Polysulfone-Sulfonated Poly(ester ester ketone) Blend Ultrafiltration Membranes. Part I. *Appl. Surface Sci., 253(21),* 8705–8712.
433. Xing Peixiang, Robertson Gilles, P., Guiver Michael, D., Mikhailenko Serguei, D., & Kaliaguine Serge. (2004). Sulfonated Poly(aryl ester ketone)s Containing Naphthalene Moieties for Proton Exchange Membranes. *J. Polym. Sci. A, 42(12),* 2866–2876.
434. Khasbulatova, Z. S., & Shustov, G. B. (2009). *Aromatic Oligomers for Synthesing Polyesters.* In Proceedings of the X International Conference of Chemistry and Physicochemistry of Oligomers. Volgograd, 99. [in Russian].
435. Khasbulatova, Z. S., Shustov, G. B., & Mikitaeva, A. K. (2010). *Vysokomol. Soed. B, 52(4),* 702–705.

CHAPTER 8

ON THERMO-ELASTOPLASTIC PROPERTIES: A CASE STUDY

MARIA RAJKIEWICZ, MARCIN ŚLĄCZKA, and JAKUB CZAKAJ

CONTENTS

ABSTRACT

The structure and physical properties of the thermoplastic vulcanizates (TPE-V) produced in the process of the reactive processing of polypropylene (PP) and ethylene-octene elastomer (EOE) in the form of alloy, using the cross-linking system was analyzed. With the DMTA, SEM and DSC it has been demonstrated that the dynamically produced vulcanizates constitute a typical dispersoid, where semicrystal PP produces a continuous phase, and the dispersed phase consists of molecules of the cross-linked ethylene-octene elastomer, which play a role of a modifier of the properties and a stabilizer of the two-phase structure. It has been found that the mechanical as well as the thermal properties depend on the content of the elastomer in the blends, exposed to mechanical strain and temperature. The best results have been achieved for grafted/cross-linked blends with the contents of iPP/EOE-55/45%.

8.1 INTRODUCTION

The structure and physical properties of the thermoplastic vulcanizates (TPE-V) produced in the process of the reactive processing of polypropylene (PP) and ethylene-octene elastomer (EOE) in the form of alloy, using the cross-linking system was analyzed. With the DMTA, SEM and DSC it has been demonstrated that the dynamically produced vulcanizates constitute a typical dispersoid, where semicrystal PP produces a continuous phase, and the dispersed phase consists of molecules of the cross-linked ethylene-octene elastomer, which play a role of a modifier of the properties and a stabilizer of the two-phase structure. It has been found that the mechanical as well as the thermal properties depend on the content of the elastomer in the blends, exposed to mechanical strain and temperature. The best results have been achieved for grafted/cross-linked blends with the contents of iPP/EOE-55/45%.

The thermoplastic elastomers (TPE) are a new class of the polymeric materials, which combine the properties of the chemically cross-linked rubbers and easiness of processing and recycling of the thermoplastics [1–8]. The characteristics of the TPE are phase micrononuniformity and specific domain morphology. Their properties are intermediate and are in the range between those, which characterize the polymers, which produce the rigid and elastic phase. These properties of TPE, regardless of its type and structure, are a function of its type, structure and content of both phases, nature and value of interphase actions and manner the phases are linked in the system.

The progress in the area of TPE is connected with the research oriented to improve thermal stability of the rigid phase (higher T_g) and to increase chemical resistance as well as thermal and thermo-oxidative stability of the elastic phase [2]. A specific group among the TPE described in the literature and used in the technology are microheterogeneous mixes of rubbers and plastomers, where the plastomer constitutes a continuous phase and the molecules of rubber dispersed in it are cross-linked during a dynamic vulcanizing process. The dynamic vulcanization is conducted during the reactive mixing of rubber and thermoplast in the smelted state, in conditions of action of variable coagulating and stretching stresses and of high coagulating speed, caused by operating unit of the equipment. Manner of producing the mixtures and their properties as well as morphological traits made them be called thermoplastic vulcanizates (TPE-V) [9, 10]. They are a group of "customizable materials" with configurable properties. To their advantage is that most of them can be produced in standard equipment for processing synthetics and rubbers, using the already available generations of rubbers as well as generations of rubbers newly introduced to the market with improved properties. A requirement of developing the system morphology (a dispersion of macromolecules of the cross-linked rubber with optimum size in the continuous phase of plastomer) and achieving an appropriate thermoplasticity necessary for TPE-V are the carefully selected conditions of preparing the mixture, (temperature, coagulation speed, type of equipment, type and amount of the cross-linking substance). When selecting type and content of elastomer, the properties of the newly created material can be adjusted toward the desirable direction. Presence of the cross-linked elastomer phase allows for avoiding glutinous flow under the load, what means better elasticity and less permanent distortion when squeezing and stretching the material produced in such a manner as compared to the traditional mixtures prepared from identical input materials, each of which produces its own continuous phase. With the dynamic vulcanization process many new materials with configured properties have been achieved and introduced to the market. The most important group of TPE-V, which has commercial significance, are products of the dynamic vulcanization of isotactic polypropylene (iPP) and ethylene-propylene-diene elastomer (EPDM). It is a result of properties of the PP and EPDM system, which, due to presence of the double bonds, may be cross-linked with conventional systems, which are of relatively low price, good contents miscibility and ability to be used within the temperature range 233–408 –K [11–15].

Next level in the field of thermoplastic elastomers began with development of the metallocene catalysts and their use in stereo block polymerization of ethylene and propylene and copolymerization of these olefins with other monomers, leading to macromolecules with a "customized" structure with a microstructure and stereo regularity defined upfront. The catalysts enabled production of the homogeneous olefin copolymers, which have narrow distribution of molecular weights (RCC=M_w/M_n<2.5). According to the developed technology called Incite and using the on-place catalyst [16, 17]. The Dow Chemical Co company produces the olefin elastomers Engage™ which contain over 8% of octene. Co-polymers Engage are characterized by lack of relation between the traditional Mooney viscosity and the technological properties. Compared to other homogeneous polymers with the same flow index, they are characterized by higher dynamic viscosity at zero coagulation speed and decreasing viscosity at increasing coagulation speed. They have no fixed yield point. Saturated nature of elastomer, caused by absence of diene in the chain, results in some restrictions in choice of the cross-linking system. The ethylene-octene elastomers can be easily radiation cross-linked with peroxides or moisture, if they are formerly grafted with silanes. There is relatively not much description of the behavior of the ethylene-octene elastomer in the dynamic vulcanization process in the literature. The specific physical properties of such elastomers and possibility to process them within a periodic process as well as within a continuous process, due to convenient form of the commercial product (granulate) encouraged us to start the recognition works on development of the technology of producing thermoplastic vulcanizates from the mixture of elastomer Engage and iPP.

Use a silane-based cross-linking system in the dynamic vulcanization process seemed the most interesting. For the research works one of the known methods of cross-linking poliolefins with silanes was used, assuming that the cross-linking of EOE would proceed according to the analogous mechanism. In the seventies of the twentieth century the Dow Corning Co. company developed two methods of the hydrolytic cross-linking of polyolefins grafted with vinylosilanes according to the radical mechanism [18, 19]. Nowadays three polyolefin cross-linking methods are widely used in the industrial production. The grounds for distinguishing them are technological equipment and procedure. It is one-phase and two-phase method and a "dry silane method," available only under license [20]. The mechanism of cross-linking PE with the cross-linking system: silane/peroxide/moisture is shown schematically in Fig. 8.1.

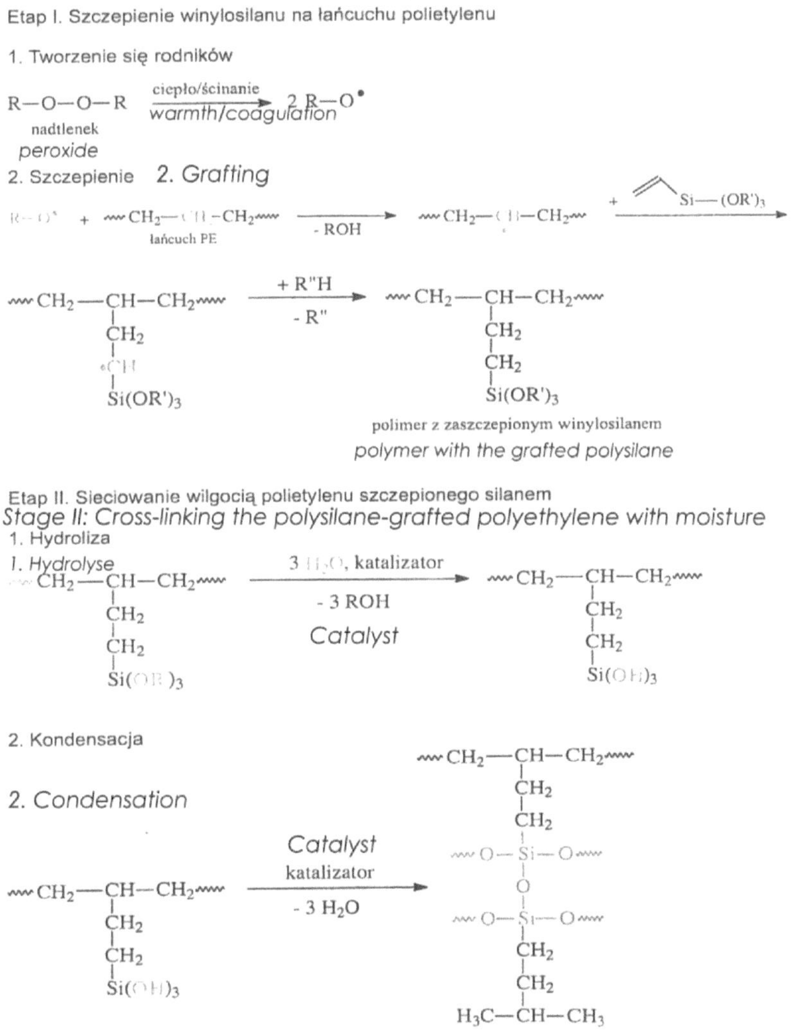

FIGURE 8.1 Crosslinking mechanism of polyolefins with a silane.

The process of catalytic hydrolyzes of the alcoxylene groups of the grafted silane to the silane groups and, then, the catalytic condensation of the silane groups leads to production of the cross-linked structure through the siloxane groups. The hydrolyze and the condensation take place in an increased temperature with presence of the catalyst and water. Dibutyl tin dilaurate (DBTL) is most often used as a catalyst of the reaction. The catalyst may be added either to the polymeric blend (it constitutes an increased risk of the premature cross-linking), or in the form of a premixed reagent during the processing. The mechanism of action of DBTL, as a cross-linking catalyst, is complex and has not been sufficiently explained.

As a result of cross-linking the polyolefins with silanes the Si-O-Si bonds are produced, which are more elastic than the rigid bonds C-C created as a result of cross-linking of polymers induced by radiation and peroxides (Fig. 8.2). Use of silanes gives more elastic products and the cross-linking process is more cost-effective.

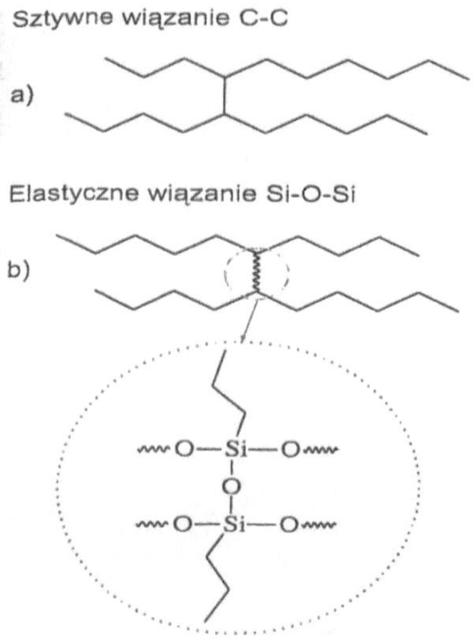

FIGURE 8.2 Structure of polyolefin's cross-linked with a (a) peroxide or radiation (b) with a silane.

8.2 EXPERIMENTAL PART

8.2.1 RAW MATERIALS

- Isotactic polypropylene Malen P-F401 iPP, for extrusion, made by Orlen SA; flow index 2.4–33.2 g/10 min, yield point in stretching 28.4 MPa, crystallinity level 95%.
- The ethylene-n-octene elastomer s EOE type Engage, synthesized according to the Insite technological process, manufactured by Du-Pont Dow Chemical Elastomers (Table 8.1).
- Silanes: Silquest A-172 vinylo-tris (2-methoxyethoxysilane), Silquest A-174–3-methacryloxypropyltrimethoxysilane, manufactured by Vitco SA.
- Dicumyl peroxide with 99% content of the neat peroxide, manufactured by ELF Atochem.
- Antioxidant tetra-kis (3,5-di-tetra-butyl-4-hydroxyphenyl) propionate, manufactured Ciba-Geigy.

8.2.2 TEST METHOD

Three types of EOE from the wide range offered by the manufacturer were selected (Table 8.1). The general-purpose elastomers were selected with high content of octene and a defined characteristic.

TABLE 8.1 Properties of Ethylene-Octene Elastomers Engage

Elastomer type Properties	Engage I	Engage II	Engage III
Co-monomer content, % of weight (^{13}C NMR/FTR)	42	40	38
Density, g/cm^3, ASTM	0.863	0.868	0.870
Mooney viscosity, ML (1+4) 121 °C	35	35	8
MFR, deg/min, ASTM D-1238	0.5	0.5	5.0
Shore hardness A, ASTM D-2240	66	75	75

The test were made aimed for determining a threshold value of content of elastomer in the iPP/EOE mixture, which was subject of the dynamic vulcanization process, considering the influence of these parameters on variable properties of iPP. A series of tests was made, in which the proportions of PP and elastomer Engage I were changed in the range 15–60%, with continuous addition of the cross-linking system (silane A-172/ dicumyl peroxide) 3/0.03% in relation to elastomer and antioxidant additive 0.2%.

For the tests of preparing dynamic vulcanizates in the continuous process of reactive extrusion a twin-screw mixer-extruder DSK 42/6D manufactured by Brabender was used. The vulcanizates were produced dynamically in the process of one-stage or two-stage extrusion process, setting the favorable operating parameters for the device, which had been determined based on multiple tests: distribution of temperatures in each heating area of the extruder: 170/180/190 °C, screw rotation: 40/min. In the one-stage process all the components provided in the formula (elastomer, iPP, antioxidant, silane initiating system/peroxide) was initially mixed in a fast-rotating mixer type Stephan in temperature of 50 °C, next a granulate was extruded. In the two-stage process in the first stage the iPP, elastomer and antioxidant mixture was extruded, next after mixing the granulates with the cross-linking system it was extruded again.

The profiles were formed from the granulates with an injection molding machine type ARBURG-420 M1000–25 All rounder. For the tests the actual injection at speed of 10 cm^3/s was used with addition at speed of 15 cm^3/s, injection temperatures: 195/200/210/210 °C and blend injection time was slightly lower than for iPP itself.

8.2.2.1 METHODS OF ANALYZING THE BLEND

Hardness was marked according to the Shore method, scale D according to PN-ISO 868 or according to the ball insertion method according to the PN-ISO 868 (MPa). Flow speed index (MFR) was determined according to PN-ISO 1033. Resistance properties of the blend with static stretching were tested according to ISO-527, using a digital tester Instron 4505 (tear off speed: 50 mm/min). The bending properties were determined according to PN-EN ISO 178. In addition to the regular tests, the selected blends were subject to specialist examination, such as the thermo gravimetric analysis (TGA), electron microscopy (SEM), differential scanning calorimetry (DSC) and dynamic thermal analysis of mechanical properties (DMTA).

The samples were heated in the ambient temperature in temperature range of 30–490 °C with speed of 5 deg/min. The test was conducted with thermo balance TGA manufactured by "Perkin Elmer." Turning points were made after freezing the samples in the liquid nitrogen for about 3 min. The surfaces of the turning points were concocted with gold with vacuum powdering. The scanning electron microscope JSM 6100 manufactured by JEOL was used to conduct the tests. The photographs have been made in magnification of 2000x.

8.3 RESULTS OF THE TESTING

Influence of the content of elastomer Engage I on physical properties of TPE with PP and EOE modified (grafting/cross-linking) with a silane/peroxide cross-linking system has been shown in Fig. 8.3 and Table 8.2. The content of comonomer had significant influence on such properties of the elastomer as elasticity, modulus, density and hardness. Values of two last parameters decreased with the increase of content of *n*-octene in elastomer. It has been stated that properties of the dynamically vulcanized blends could be adjusted with content of the elastomer phase. With the increase of content of EOE in range 15–45% tensile strength increased (18–30 MPa), and, in the same time, relative elongation increased with tear off (300–700%). With elastomer content over 50% a visible decrease of both properties occurred, which came to 15 MPa and 600%, respectively. Whereas hardness expressed in Sh degrees or in MPa) systematically decreased with the increase of the content of EOE in the blend. The optimum content of EOE introduced to PP was 45% and therefore in most subsequent tests a blend was used, in which iPP/EOE ratio was 55/45%. Such contents had also the blends listed in Table 8.3, made of three types of EOE and two types of silane, with constant content of the cross-linking system (silane A–174/dicumyl peroxide 3.0/0.01%, irganox 1010–0.2%. The blends containing elastomers Engage I and Engage II, with difference of content of octene by 2% Shore hardness A (66 and 75, respectively) and with very similar Mooney viscosities, showed comparable resistance and rheological characteristic. The blends containing elastomer Engage

III, with the lowest octene content, were characterized by slightly lower variables of tensile strength (tension at the tear off), elongation and hardness, but by a much higher tension at the yield point and high flow index.

TABLE 8.2 Selected Properties of the Dynamically Cross Linked Blends in Relations To PP/EOE Ratio

Ratio PP/Engage I, % of weight	100/0	85/15	70/30	55/45	40/60
Hardness, °ShD	80	63	57	50	36
MFR (190 °C, 2.16 kg), g/10 min	2.4	1.63	1.29	1.28	1.15
MFR (190 °C, 5), g/10 min	—	5.06	5.89	5.80	4.90
$T_{A\,120}$, °C*	152	143	130	106	~60
Hardness HK, MPa**	24.7	16.1	12.2	11.8	8.7
Solubility of elastomer in cyclohexane, %	—	—	13.9	12.03	14.2
Solubility of elastomer <t4/> in boiling xylene, %	—	—	24.0	33.0	42.0

*T_{A120} – Vicat softening point; **HK – ball pan hardness method.

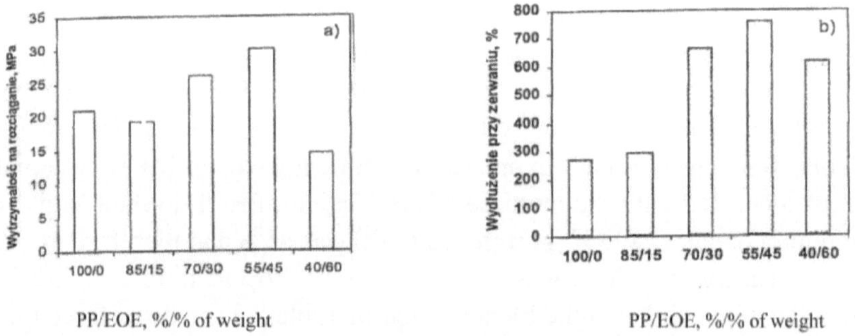

FIGURE 8.3 Mechanical properties of the dynamically cross linked blends in relation to PP/EOE ratio, (a) tensile strength, (b) elongation at the tear off.

TABLE 8.3 Effect of Type of Elastomer Engage Modified With Silane A-174 on the Properties of the Dynamically Cross-Linked PP/EOE–55/45% Blends

Elastomer	Engage I	Engage II	Engage III
Blend properties			
MFR, g/10 min			
(2.16 kg, 190 °C)	1.86	1.80	4.07
Gardbess, °Sh, D	42/39	42/40	39/38
$\epsilon_{B,}$ %	720	752	660
$\sigma_{M,}$ MPa	25.2	29.5	21.9
$\sigma_{100\%,}$ MPa	11.1	12.6	11.0
σ_{y}, MPa	11.1	12.6	11.1
ϵ_{y}, %	24.0	27.9	39.9

Symbols; $\epsilon_{100\%}$ – tension at 100% elongation, σ_M – maximum tension, ϵ_B – relative elongation at the tear off, σ_y – yield point, ϵ_y – elongation on yield point.

Such behavior of elastomer Engage III blended with PP resulted probably from its different rheological characteristic, including its four times lower viscosity and very high flow index as compared to Engage I, which was recognized as the most suitable for production of nonsaturated blends using the dynamic vulcanization method.

Blend with the selected optimum contents iPP/Engage I 55/45% and the selected cross-linking system (silane/peroxide 3/0.03%) were characterized by high thermal stability, independent from type of the material employed to cross-linking silane. It has been confirmed with tests of TGA of blend containing silane A-172 and silane A-174 (samples PL-1 and PL-2, respectively), what is shown in Table 8.4 in temperature of 230 °C the decrease of weight did not exceed 0.5%.

TABLE 8.4 Results of the Thermogravimetric Analysis of Ipp/EOE-55/45% Blend (Engage I)

PL-1(Silane A-172)		PL-2 (Silane A-174)	
Temperature, °C	**Decrease of weight, %**	**Temperature, °C**	**Decrease of weight, %**
230	0.23	230	0.36
300	7.43	300	7.66
352	25.97	378	54.36
363	40.06	405	89.63
430	94.05	426	94.36

In temperature 300 °C came to as much as 7.5%, and the further increase of temperature caused the progressive degradation process. Analysis of the morphological structure of grafted/cross-linked iPP/Engage I blend using the SEM, DSC and DMTA methods showed that the blends produced with dynamic vulcanization had a special two-phase structure. With scanning electron microscopy photographs of surface of turning points of iPP samples and iPP/Engage I 55/45% blends have been made (Fig. 8.4). The SEM analysis showed that the obtained blends were mixtures of two thermodynamically nonmiscible structures. The continuous phase of iPP had a visible semicrystal structure and the spherical and oval molecules of the dispersed properties, assessed with the DMTA Mk II equipment manufactured by Polymer Laboratories in the sinusoidally variable load conditions at bending with frequency of 1 Hz, in the temperature range between −100 and +100 °C also showed heterogeneous structure of the produced blends.

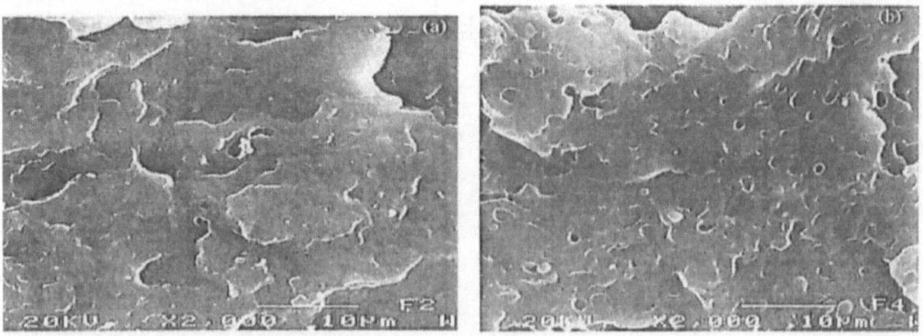

FIGURE 8.4 SEM microphotographs of: (a) neat iPP, (b) dynamically cross-linked PP/EOE blend 55/45%; magnification 2000x.

In Fig. 8.5, course of changes of the storage modulus E,' loss modulus E" and vibration damping factor gδ for iPP and iPP/Engage I blends with content of 85/15, 70/30 and 55/45% in relation to temperature has been shown. For iPP/EOE blends two, clear relaxation transitions in the range of glass transition are visible, near glass points of iPP and EOE.

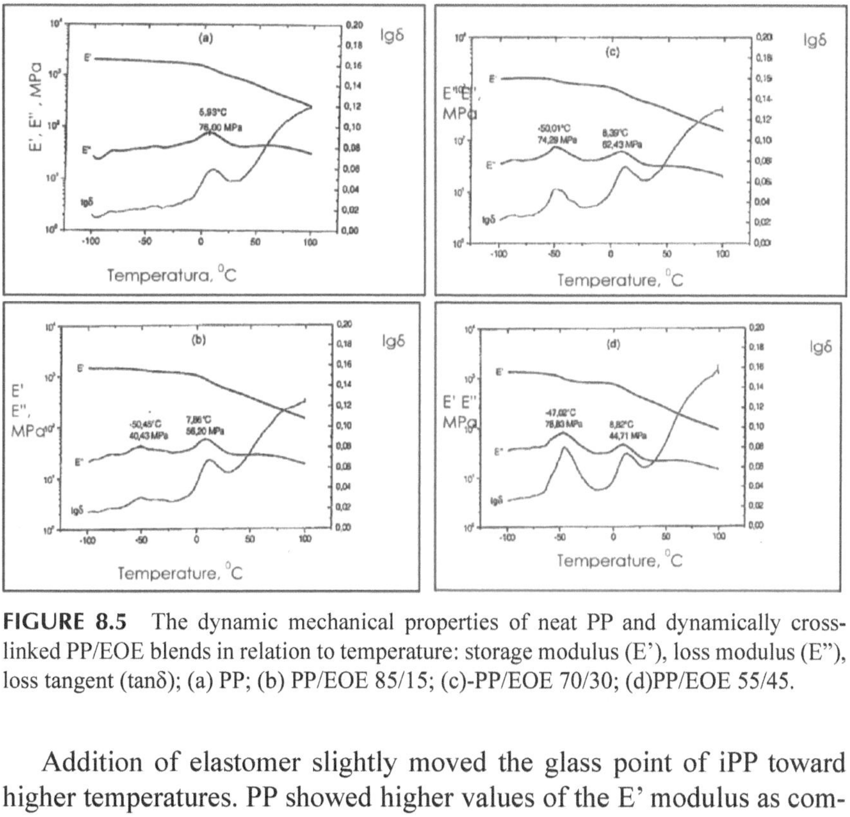

FIGURE 8.5 The dynamic mechanical properties of neat PP and dynamically cross-linked PP/EOE blends in relation to temperature: storage modulus (E'), loss modulus (E"), loss tangent (tanδ); (a) PP; (b) PP/EOE 85/15; (c)-PP/EOE 70/30; (d)PP/EOE 55/45.

Addition of elastomer slightly moved the glass point of iPP toward higher temperatures. PP showed higher values of the E' modulus as compared to the analyzed composites, whereas in the chart E" one maximum appeared corresponding to T_g PP.

On the DSC thermal images made in positive temperatures (50–210 °C) a visible maximum appeared, which was connected with thermal transition corresponding to the melting point of iPP. Systematic decrease of melting point of the thermoplastic phase of iPP in iPP/Engage I blends related to the original polymer was observed (Fig. 8.6). Causes of these changes could not be unambiguously determined it is supposed that here the phenomena such, as degradation of iPP in conditions of the high-temperature processing change of semicrystal structure of iPP may have occurred. In order to compare the properties of iPP/Engage I blend with content of 55/45%, produced with periodical method and in the one-stage or two-stage continuous process, a series of tests with use of the general formula was performed. Properties of the selected blends cross-linked with the silane A-174/dicumyl peroxide (TE-1) and silane A-172/dicumyl peroxide (TE-2) systems have been listed in Table 8.5.

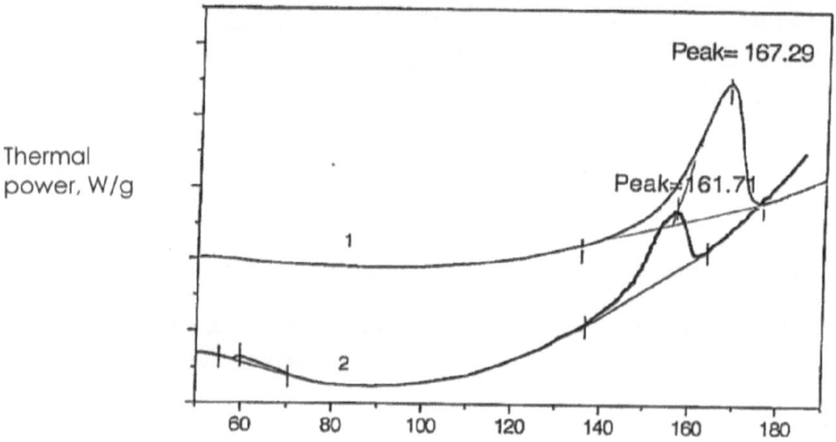

FIGURE 8.6 DSC curves of (1) PP, (2) PP/EOE dynamically cross-linked 55/45%.

TABLE 8.5 Selected Properties of PP and Dynamically Cross-Linked PP/EOE Blend PP/EOE (55/45%) in the Reactive Processing

Properties	PP	TE-1 (Silane A-174)	TE-2 (Silane A-172
Yield point, MPa	31.4	12.1	12.5
Relative elongation of yield point, %	9.4	25.1	33.0
Tensile strength, MPa	15.1	15.5	18.3
Elongation at the tear off, %	196	443	440

TABLE 8.5 *(Continued)*

Properties	**PP**	**TE-1 (Silane A-174)**	**TE-2 (Silane A-172**
Properties	PP	TE-1 (Silane A-174)	TE-2 (Silane A-172
Tensile shear modulus, MPa	1455	476	430
Bending strength, MPa	39.6	13.0	10.4
Tensile bending modulus, MPa	1499	504	415
Young modulus, MPa	1520	515	502
HDT, load 1.8 MPa, °C	50.5	38	38
Izod notched impact strength, kJ/m^2	3.16	46.7	43.0

8.4 CONCLUSIONS

The conducted tests leaded to developing grounds for the technology for dynamic vulcanization of materials with thermo-elastoplastic properties, in which a thermoplastic polymer constitutes a continuous phase, whereas the dispersed phase consists of cross-linked elastomer particles. Basic elastomers are polyisoprene with isotactivity level of 85% or higher and copolymer EOE containing over 30% of n-octene.

The blend properties can be adjusted with content of the elastomeric phase and with cross-linking silane/peroxide system. The achieved material has a heterogeneous structure and favorable set of resistance properties, including higher Izod notched impact strength as compared to PP.

The developed technology allows for achieving new materials with preconfigured properties, competitive to the unmodified iPP and to physical PP/elastomer mixtures. They could be processed in the equipment for synthetics. These blends may be used as structural materials, characterized by higher thermal resistance as compared to the unmodified PP. They may also be a generation of modifiers for polyolefins and polymeric mixtures.

KEYWORDS

- **Ethylene-octane**
- **Polypropylene**
- **Thermoplastic composites**

REFERENCES

1. Rzymski, W., & Radusch, H. J. (2002). *Polimery, 47*, 229.
2. Spontak, J. Richard, & Patel Nikunj, P. (2000). Current Opinion in Colloids and Interface. *Science*, *5*, 333.
3. Rzymski, W., & Radusch, H. J. (2005). *Polimery*, *50*, 247.
4. Radusch, H. J., Dosher P., & Lohse G. (2005). *Polimery*, *50*, 279.
5. Rzymski, W. M. (1998)., *Stosowanie i Przetwórstwo Materiałów Polimerowych.* Wyd. Polit. Częstochowskiej, Częstochowa s. 17–28.
6. Holden, *G.* (2000). *Understanding Thermoplastic Elastomers,* Hanser Publishers. Munich.
7. Rader, C. P. (1991). *Modern Plastic Encyclopedia.*
8. Rader, C. P. (1993). *Kuststoffe*, *83*, 777.
9. Rzymski, W., & Radusch, H. J. (2001). *Elastomery*, *5(2)*, 19.
10. Rzymski, W., & Radusch, H. J. (2001). *Elastomery*, *5(3)*, 3.
11. Winters, R. R. (2001). *Polimery*, *42*, 9745.
12. Trinh, An Huy., Luepke, T., & Radusch, H. J. (2001). *App. Polym. Sci.*, *80*, 148.
13. Jain, A. K., Nagpal, A. K., Singhal, R., & Gupta Neeraj, K. (2000). *J. Appl. Polym. Sci.* *78*, 2089.
14. Gupta Neeraj, K., Janil Anil, K., Singhal, R., & Nagpal, A. K. (2000). *J. Appl. Polym. Sci.*, *78*, 2104.
15. Suresh, S. Chandra Kumar, Alagar, M. Anand Prabu, A. (2003). *Eur. Polym. J.*, *39*, 805.
16. Fanicher, L., & Clayfield, T. (1997). *Elastomery*, *1(4)*.
17. ENGAGE – Polyolefin Elastomers, A Product of Du Pont Dow Elastomers. Product Information, 2003.
18. Voight, H. U. (1981). *Kautsch. Gum. Kunstst*, *34*, 197.
19. Toynbee, J. (1994). *Polymer*, *35*, 428.
20. *Special Chem.* Crosslinking Agent Center, Dane Techniczne, 2004.

CHAPTER 9

MODELING, SIMULATION, PERFORMANCE AND EVALUATION OF CARBON NANOTUBE/POLYMER COMPOSITES

A. K. HAGHI and G. E. ZAIKOV

CONTENTS

ABSTRACT

In this chapter, new trends in computational chemistry and computational mechanics for the prediction of the structure and properties of CNT materials are presented simultaneously.

9.1 INTRODUCTION

It has been known that the mechanical properties of polymeric materials like stiffness and strength can be engineered by producing composites that are composed of different volume fraction of one or more reinforcing phases. In traditional form, polymeric materials have been reinforced with carbon, glass, basalt, ceramic and aramid microfibers to improve their mechanical properties. These composite materials have been used in many applications in automotive, aerospace and mass transit. As time has proceeded, a practical accomplishment of such composites has begun to change from micro scale composites to nanocomposite, taking advantages of better mechanical properties. While some credit can be attributed to the intrinsic properties of the fillers, most of these advantages stem from the extreme reduction in filler size combined with the large enhancement in the specific surface area and interfacial area they present to the matrix phase. In addition, since traditional composites use over 40 wt.% of the reinforcing phase, the dispersion of just a few weight percentages of nanofillers into polymeric matrices could lead to dramatic changes in their mechanical properties. One of the earliest nanofiller that witch have received significant and shown super mechanical properties is Carbon Nano Tube (CNT), because of their unique properties, CNTs have a wide range of potentials for engineering applications due to their exceptional mechanical, physical, electrical properties and geometrical characteristics consisting of small diameter and high aspect ratio. It has shown that dispersion of a few weight percentages of nanotubes in a matrix dramatically increase mechanical, thermal and electrical properties of composite materials. Development of CNT-nanocomposites requires a good understanding of

CNT's and CNT's nanocomposite properties. Because of the huge cost and technological difficulties associated with experimental analysis at the scale of nano, researchers are encouraged to employ computational methods for simulating the behavior of nanostructures like CNTs from different mechanical points of view.

To promote the design and development of CNT-nanocomposite materials, structure and property relationships must be recognized that predict the bulk mechanical of these materials as a function of the molecular and micro structure mechanical properties of nanostructured materials can be calculated by a select set of computational methods. These modeling methods extend across a wide range of length and time scales, as shown in Fig. 9.1. For the smallest length and time scales, a complete understanding of the behavior of materials requires theoretical and computational tools that span the atomic-scale detail of first principles methods (density functional theory, molecular dynamics, and Monte Carlo methods). For the largest length and time scales, computational mechanics is used to predict the mechanical behavior of materials and engineering structures. And the coarser grained description provided by continuum equations. However, the intermediate length and time scales do not have general modeling methods that are as well developed as those on the smallest and largest time and length scales. Therefore, recent efforts have focused on combining traditional methodologies and continuum descriptions within a unified multi scale framework. Multi scale modeling techniques are employed, which take advantage of computational chemistry and computational mechanics methods simultaneously for the prediction of the structure and properties of materials.

As illustrated in Fig. 9.1, in each modeling method has extended classes of related modeling tools that are shown in a short view in a diagram in Fig. 9.2.

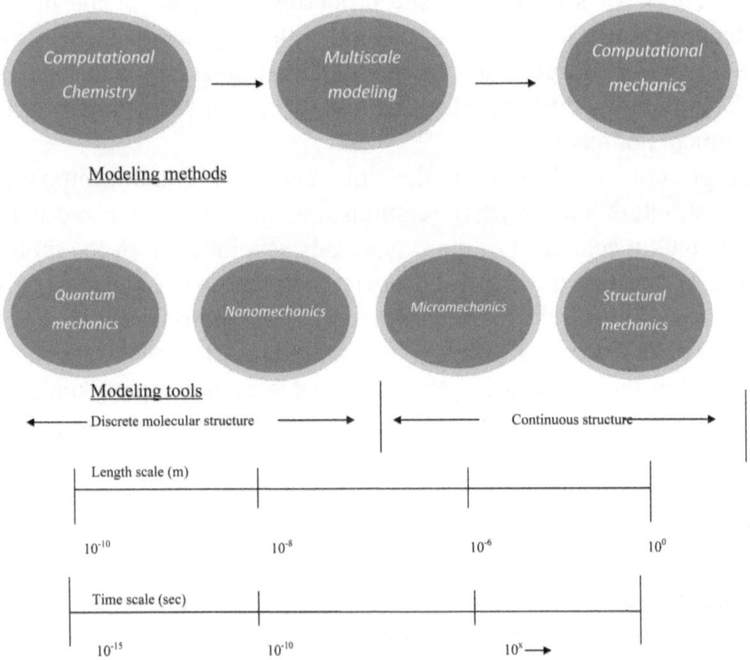

FIGURE 9.1 Different length and time scale used in determination mechanical properties of polymer nano-composite.

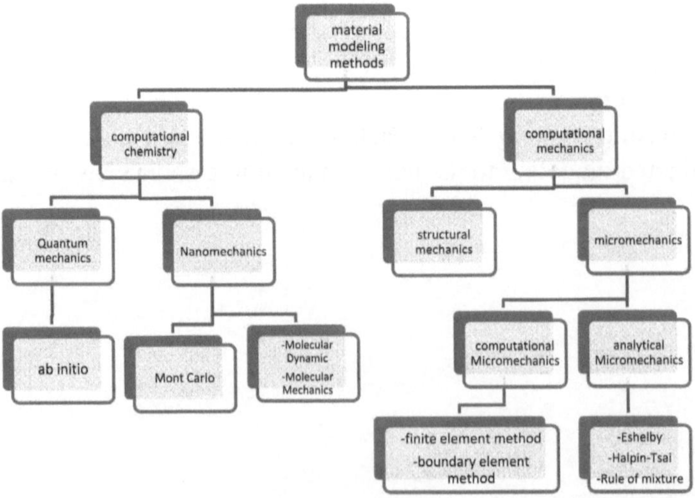

FIGURE 9.2 Diagram of material modeling methods.

9.2 MATERIAL MODELING METHODS

There are different of modeling methods currently used by the researches. They plan not only to simulate material behavior at a particular scale of interest but also to assist in developing new materials with highly desirable properties. These scales can range from the basic atomistic to the much coarser continuum level. The hierarchy of the modeling methods consist quantum mechanics, molecular dynamic, micromechanics and finally continuum mechanics, that could be categorized in tow main group: atomistic modeling and computational mechanics.

9.2.1 COMPUTATIONAL CHEMISTRY (ATOMISTIC MODELING)

The atomistic methods usually employ atoms, molecules or their group and can be classified into three main categories, namely the quantum mechanics (QM), molecular dynamics (MD) and Monte Carlo (MC). Other atomistic modeling techniques such as tight bonding molecular dynamics (TBMD), local density (LD), dissipative particle dynamics (DPD), lattice Boltzmann (LB), Brownian dynamics (BD), time-dependent Ginzburg–Lanau method, Morse potential function model, and modified Morse potential function model were also applied afterwards.

9.2.1.1 QUANTUM MECHANICS

The observable properties of solid materials are governed by quantum mechanics, as expressed by solutions of a Schrödinger equation for the motion of the electrons and the nuclei. However, because of the inherent difficulty of obtaining even coarsely approximate solutions of the full many body Schrödinger equation, one typically focuses on reduced descriptions that are believed to capture the essential energetic of the problem of Interest Tow main quantum mechanics method are "Ab initio" and "density function method (DFT)."

Unlike most materials simulation methods that are based on classical potentials, the main advantages of Ab initio methods, which is based on first principles density functional theory (without any adjustable parameters), are the generality, reliability, and accuracy of these methods. They involve the solution of Schrödinger's equation for each electron, in the self-consistent potential created by the other electrons and the nuclei. Ab initio methods can be applied to a wide range of systems and properties. However, these techniques are computationally exhaustive, making them difficult for simulations involving large numbers of atoms. There are three widely used procedures in Ab initio simulation. These procedures are single point calculations, geometry optimization, and frequency calculation. Single point calculations involve the determination of energy and wave functions for a given geometry. This is often used as a preliminary step in a more detailed simulation. Geometry calculations are used to determine energy and wave functions for an initial geometry, and subsequent geometries with lower energy levels. A number of procedures exist for establishing geometries at each calculation step. Frequency calculations are used to predict Infrared and Raman intensities of a molecular system. Ab initio simulations are restricted to small numbers of atoms because of the intense computational resources that are required.

Ab initio techniques have been used on a limited basis for the prediction of mechanical properties of polymer based nano structured composites.

9.2.1.2 MOLECULAR DYNAMIC

MD is a simulation technique that used to estimate the time depended physical properties of a system of interacting particles (e.g., atoms, molecules, etc.) by predict the time evolution. MD simulation used to investigating the structure, dynamics, and thermodynamics of individual molecules.

There are two basic assumptions made in standard molecular dynamics simulations.

Molecules or atoms are described as a system of interacting material points, whose motion is described dynamically with a vector of instantaneous positions and velocities. The atomic interaction has a strong dependence on the spatial orientation and distances between separate atoms. This model is often referred to as the soft sphere model, where the softness is analogous to the electron clouds of atoms.

No mass changes in the system. Equivalently, the number of atoms in the system remains the same.

The atomic position, velocities, and accelerations of individual particles that vary with time. The described by track that MD simulation generates and then used to obtain average value of system such as energy, pressure and temperature.

The main three parts of MD simulation are:

- initial conditions (e.g., initial positions and velocities of all particles in the system).
- The interaction potentials between particles to represent the forces among all the particles.
- The evolution of the system in time by solving of classical Newtonian equations of motion for all particles in the system.

The equation of motion is generally given by Eq. (1).

$$\vec{F_i}(t) = m_i \frac{d^2\vec{r_i}}{dt^2} \ (1) \qquad (1)$$

Where is the force acting on the *i*-th atom or particle at time *t* and is obtained as the negative gradient of the interaction potential *U*, m_i is the atomic mass and the atomic position. The interaction potentials together with their parameters, describe how the particles in a system interact with each other (so-called force field). Force field may be obtained by quantum method (e.g., Ab initio), empirical method (e.g., Lennard–Jones, Mores, and Born-Mayer) or quantum-empirical method (e.g., embedded atom model, glue model, bond order potential).

The criteria for selecting a force field include the accuracy, transferability and computational speed. A typical interaction potential *U* may consist of a number of bonded and nonbonded interaction terms:

$$U(\vec{r_1}, \vec{r_2}, \vec{r_3}, \ldots, \vec{r_n})$$

$$= \sum_{i_{bond}}^{N_{bond}} U_{bond}(i_{bond}, \overrightarrow{r_a}, \overrightarrow{r_b}) +$$
$$\sum_{i_{angle}}^{N_{angle}} U_{angle}(i_{angle}, \overrightarrow{r_a}, \overrightarrow{r_b}, \overrightarrow{r_c}) + \sum_{i_{torsion}}^{N_{torsion}} U_{torsion}(i_{torsion}, \overrightarrow{r_a}, \overrightarrow{r_b}, \overrightarrow{r_c}, \overrightarrow{r_d})$$
$$+ \sum_{i_{inversion}}^{N_{inversion}} U_{inversion}(i_{inversion}, \overrightarrow{r_a}, \overrightarrow{r_b}, \overrightarrow{r_c}, \overrightarrow{r_d})$$
$$+ \sum_{i=1}^{N-1} \sum_{j>i}^{N} U_{vdw}(i, j, \overrightarrow{r_a}, \overrightarrow{r_b})$$
$$+ \sum_{i=1}^{N-1} \sum_{j>i}^{N} U_{electrostatic}(i, j, \overrightarrow{r_a}, \overrightarrow{r_b}) \quad (2)$$

$$\vec{r}(t + \delta t) = \vec{r}(t) + \vec{v}(t)\delta t + \frac{1}{2}\vec{a}(t)\delta^2 t + \cdots \quad (2)$$

The first four terms represent bonded interactions, that is, bond stretching U_{bond}, bond-angle bend U_{angle}, dihedral angle torsion $U_{torsion}$ and inversion interaction.

$U_{inversion}$.Vander Waals energy U_{vdw} and electrostatic energy $U_{electrostatic}$ are nonbonded interactions. In the equation,are the positions of the atoms or particles specifically involved in a given interaction;,, and illustrate the total numbers of interactions in the simulated system; ,, and presented an individual interaction each of them. There are many algorithms like *varlet*, *velocity varlet*, *leap-frog* and *Beeman*, for integrating the equation of motion, all of them using finite difference methods, and assume that the atomic position, velocities and accelerations can be approximated by a *Taylor series expansion*:

$$\vec{r}(t + \delta t) = \vec{r}(t) + \vec{v}(t)\delta t + \frac{1}{2}\vec{a}(t)\delta^2 t + \cdots \quad (3)$$

$$\vec{v}(t + \delta t) = \vec{v}(t) + \vec{a}(t)\delta t + \frac{1}{2}\vec{b}(t)\delta^2 t + \cdots \quad (4)$$

$$\vec{a}(t + \delta t) = \vec{a}(t) + \vec{b}(t)\delta t + \cdots \quad (5)$$

The *varlet algorithm* is probably the most widely used method. It uses the positions and accelerations at time *t*, and the positions from the previous step *(t–δ)* to calculate the new positions at *(t+δt)*, so:

$$\vec{r}(t + \delta t) = \vec{r}(t) + \vec{v}(t)\delta t + \frac{1}{2}\vec{a}(t)\delta t^2 + \cdots \quad (6)$$

$$\vec{r}(t - \delta t) = \vec{r}(t) - \vec{v}(t)\delta t + \frac{1}{2}\vec{a}(t)\delta t^2 + \cdots \quad (7)$$

$$\vec{r}(t + \delta t) = 2\vec{r}(t) - \vec{r}(t - \delta t) + \vec{a}(t)\delta t^2 + \cdots \quad (8)$$

The velocities at time *t* and can be, respectively estimated.

$$\vec{v}(t) = [\vec{r}(t + \delta t) - \vec{r}(t - \delta t)]/2\delta t \quad (9)$$

$$\vec{v}(t + 1/2\delta t) = [\vec{r}(t + \delta t) - \vec{r}(t - \delta t)]/\delta t \quad (10)$$

The advantages of this method are *time-reversible* and *good energy conservation* properties, where disadvantage is *low memory storage* because the velocities are not included in the time integration. However, removing the velocity from the integration introduces numerical inaccuracies method, namely the *velocity Verlet* and *Verlet leap-frog algorithms* as mentioned before, which clearly involve the velocity in the time evolution of the atomic coordinates.

9.2.1.3 MONTE CARLO

Monte Carlo technique (*Metropolis* method) use random number from a given probability distribution to generate a sample population of the system from which one can calculate the properties of interest. a MC simulation usually consists of three typical steps:

- The physical problem is translated into an analogous probabilistic or statistical model.
- The probabilistic model is solved by a numerical stochastic sampling experiment.
- The obtained data are analyzed by using statistical methods.

Athwart MD which provides information for nonequilibrium as well as equilibrium properties, MC gives only the information on equilibrium properties (e.g., free energy, phase equilibrium). In a NVT ensemble with N atoms, new formation the change in the system *Hamiltonian* (ΔH) by randomly or systematically moving one atom from position $i \rightarrow j$ can calculated.

$$\Delta H = H(j) - H(i) \tag{11}$$

Where *H(i)* and *H(j)* are the *Hamiltonian* associated with the original and new configuration, respectively.

The $\Delta H<0$ shows the state of lower energy for system. So, the movement is accepted and the new position is stable place for atom.

For $\Delta H \geq 0$ the move to new position is accepted only with a certain probability P which is given by

$$Pi \rightarrow j \propto exp\left(-\frac{\Delta H}{K_B T}\right) \tag{12}$$

is the Boltzmann constant.

According to *Metropolis* method, a random number ζ between 0 and 1 could be generated and determine the new configuration according to the following rule:

$$\text{For } \zeta \leq exp\left(-\frac{\Delta H}{K_B T}\right); \text{ the move is accepted;} \tag{13}$$

$$\text{For } \zeta > exp\left(-\frac{\Delta H}{K_B T}\right); \text{ the move is not accepted.} \tag{14}$$

If the new configuration is rejected, repeats the process by using other random chosen atoms.

In a μVT ensemble, a new configuration *j* by arbitrarily chosen and it can be exchanged by an atom of a different kind. This method affects the chemical composition of the system and the move is accepted with a certain probability. However, the energy, ΔU, will be changed by change in composition.

If $\Delta U < 0$, the move of compositional change is accepted. However, if $\Delta U \geq 0$, the move is accepted with a certain probability which is given by:

$$P_{i \to j} \propto \exp\left(-\frac{\Delta U}{K_B T}\right) \tag{15}$$

where ΔU is the change in the sum of the mixing energy and the chemical potential of the mixture. If the new configuration is rejected one counts the original configuration as a new one and repeats the process by using some other arbitrarily or systematically chosen atoms. In polymer nanocomposites, MC methods have been used to investigate the molecular structure at nano particle surface and evaluate the effects of various factors.

9.2.2 COMPUTATIONAL MECHANICS

The continuum material that assumes is continuously distributed throughout its volume have an average density and can be subjected to body forces such as gravity and surface forces. Observed macroscopic behavior is usually illustrated by ignoring the atomic and molecular structure. The basic laws for continuum model are:

- continuity, drawn from the conservation of mass;
- equilibrium, drawn from momentum considerations and Newton's second law;
- the moment of momentum principle, based on the model that the time rate of change of angular momentum with respect to an arbitrary point is equal to the resultant moment;
- conservation of energy, based on the first law of thermodynamics;
- conservation of entropy, based on the second law of thermodynamics.

These laws provide the basis for the continuum model and must be coupled with the appropriate constitutive equations and the equations of state to provide all the equations necessary for solving a continuum problem. The continuum method relates the deformation of a continuous medium to the external forces acting on the medium and the resulting internal stress and strain. Computational approaches range from simple closed-form analytical expressions to micromechanics and complex structural mechanics calculations based on beam and shell theory. In this section, we introduce some continuum methods that have been used in polymer nanocomposites, including micromechanics models (e.g., Halpin–Tsai model, Mori–Tanaka model and finite element analysis.) and the semicontinuum methods like equivalent-continuum model and will be discussed in the next part.

9.2.2.1 MICROMECHANICS

Micromechanics are a study of mechanical properties of unidirectional composites in terms of those of constituent materials. In particular, the properties to be discussed are elastic modulus, hydrothermal expansion coefficients and strengths. In discussing composites properties it is important to define a *volume element* which is small enough to show the microscopic structural details, yet large enough to present the overall behavior of the composite. Such a volume element is called the *Representative Volume Element (RVE)*. A simple representative volume element can consists of a fiber embedded in a matrix block, as shown in Fig. 9.3.

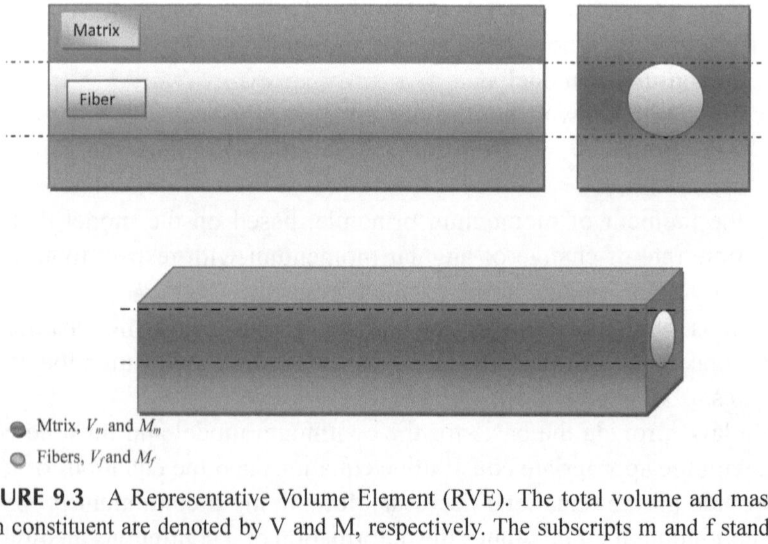

FIGURE 9.3 A Representative Volume Element (RVE). The total volume and mass of each constituent are denoted by V and M, respectively. The subscripts m and f stand for matrix and fiber, respectively.

One a representative volume element is chosen, proper boundary conditions are prescribed. Ideally, these boundary conditions must represent the in situ state of stress and strain within the composite. That is, the prescribed boundary conditions must be the same as those if the representative volume element were actually in the composite.

Finally, a prediction of composite properties follows from the solution of the foregoing boundary value problem. Although the procedure involved is conceptually simple, the actual solution is rather difficult. Consequently, many assumption and approximation have been introduces, and therefore various solution are available.

9.2.2.1.1 BASIC CONCEPTS

Micromechanics models usually used to reinforced polymer composites, based on follow basic assumptions:

1. linear elasticity of fillers and polymer matrix;
2. the reinforcement are axis-symmetric, identical in shape and size, and can be characterized by parameters such as length-to-diameter ratio (aspect ratio);
3. Perfect bonding between reinforcement and polymer interface and the ignorance of interfacial slip, reinforcement and polymer debonding or matrix cracking.

Consider a composite of mass M and volume V, illustrated schematically in Fig. 9.3, V is the volume of a *Representative Volume Element (RVE)*, since the composite is made of fibers and matrix, the mass M is the sum of the total mass M_f of fibers and mass M_m of matrix:

$$M=M_f+M_m \tag{16}$$

The Eq. (16) is valid regardless of voids which may be present. However, the composite volume V includes the volume V_v of voids so that:

$$V = V_f + V_m + V_v \tag{17}$$

Dividing Eqs. (16) and (17), leads to the following relation for the mass fraction and volume fractions:

$$m_f + m_m = 1 \tag{18}$$

$$v_f + v_m + v_v = 1 \tag{19}$$

The composite density ρ calculated as follows:

$$\rho = \frac{M}{V} = \frac{(\rho_f v_f + \rho_m v_m)}{V} = \rho_f v_f + \rho_m v_m \tag{20}$$

$$\rho = \frac{1}{m_f/\rho_f + m_m/\rho_m + v_v/\rho} \tag{21}$$

These equations can be used to determine the void fraction:

$$v_v = 1 - \rho\left(\frac{m_f}{\rho_f} + \frac{m_m}{\rho_m}\right) \tag{22}$$

The mass fraction of fibers can be measured by removing the matrix. Based on the first concept, the linear elasticity, the linear relationship between the total stress and infinitesimal strain tensors for the reinforcement and matrix as expressed by the following constitutive equations:

$$\sigma_f = C_f \varepsilon_f \tag{23}$$

$$\sigma_m = C_m \varepsilon_m \tag{24}$$

where C is the stiffness tensor.

The second concept is the average stress and strain. While the stress field and the corresponding strain field are usually nonuniform in polymer composites, the average stress and strain are then defined over the representative averaging volume V, respectively. Hypothesize the stress field in the RVE is . Then, composite stress is defined by:

$$\bar{\sigma}_i = \frac{1}{V}\int \sigma_i dv = \frac{1}{V}\left[\int_{v_f} \sigma_i dv + \int_{v_m} \sigma_i dv + \int_{v_v} \sigma_i dv\right] \tag{25}$$

$$\bar{\sigma}_{fi} = \frac{1}{V_f}\int_{v_f} \sigma_i \, dV, \bar{\sigma}_{mi} = \frac{1}{V_m}\int_{v_m} \sigma_i \, dV \tag{26}$$

Because of no stress is transmitted in the voids, in and so:

$$\bar{\sigma}_i = v_f \bar{\sigma}_{fi} + v_m \bar{\sigma}_{mi} \tag{27}$$

Where is composite average stress, is fibers average stress and is matrix average stress. Similarly to the composite stress, the composite strain is defined as the volume average strain, and is obtained as:

$$\bar{\varepsilon}_i = v_f \bar{\varepsilon}_{fi} + v_m \bar{\varepsilon}_{mi} + v_v \bar{\varepsilon}_{vi} \tag{28}$$

Despite the stress, the void in strain does not vanished; it is defined in term of the boundary displacements of the voids. So, because of the void fraction is usually negligible. Therefore, last term in Eq. (26) could be neglected and the equation corrected to:

$$\bar{\varepsilon}_i = v_f \bar{\varepsilon}_{fi} + v_m \bar{\varepsilon}_{mi} \tag{29}$$

The average *stiffness* of the composite is the tensor *C* that related the average strain to the average stress as follow equation:

$$\bar{\sigma} = C\bar{\varepsilon} \tag{30}$$

The average *compliance S* is defined in this way:

$$\bar{\varepsilon} = S\bar{\sigma} \tag{31}$$

Another important concept is the *strain concentration* and *stress concentration* tensors *A* and *B* which are basically the ratios between the average reinforcement strain or stress and the corresponding average of the composites.

$$\bar{\varepsilon}_f = A\bar{\varepsilon} \tag{32}$$

$$\bar{\sigma}_f = B\bar{\sigma} \tag{33}$$

Finally, the average composite stiffness can be calculated from the strain concentration tensor A and the reinforcement and matrix properties:

$$C = C_m + v_f(C_f - C_m)A \tag{34}$$

9.2.2.1.2 HALPIN–TSAI MODEL

Halpin-Tsai theory is used for prediction elastic modulus of unidirectional composites as function of aspect ratio. The longitudinal stiffness, E_{11} and transverse modulus, E_{22}, are expressed in the following general form:

$$\frac{E}{E_m} = \frac{1+\zeta\eta v_f}{1-\eta v_f} \tag{35}$$

Where E and E_m are modulus of composite and matrix, respectively, is fiber volume fraction and η is given by this equation:

$$\eta = \frac{\frac{E}{E_m} - 1}{\frac{E_f}{E_m} + \zeta_f} \tag{36}$$

where fiber modulus is and ζ is shape parameter that depended on reinforcement geometry and loading direction. For E_{11} calculation, ζ is equal to l/t where l is length and t is thickness of reinforcement, for E_{22}, ζ is equal to w/t where w is width of reinforcement.

For →0, the Halpin-Tsai theory converged to inversed *rule of mixture* for stiffness.

$$\frac{1}{E} = \frac{v_f}{E_f} + \frac{1-v_f}{E_m} \tag{37}$$

If →∞, the Halpin-Tsai converge to rule of mixture.

$$E = E_f v_f + E_m(1 - v_f) \tag{38}$$

9.2.2.1.3 MORI– TANAKA MODEL

The Mori-Tanaka model is uses for prediction an elastic stress field for in and around an ellipsoidal reinforcement in an infinite matrix. This method is based on Eshebly's model. Longitudinal and transverse elastic modulus, E_{11} and E_{22}, for isotropic matrix and directed spherical reinforcement are:

$$\frac{E_{11}}{E_m} = \frac{A_0}{A_0 + v_f(A_1 + 2v_0 A_2)} \tag{39}$$

$$\frac{E_{22}}{E_m} = \frac{2A_0}{2A_0 + v_f(-2A_3 + (1 - v_0 A_4) + (1 + v_0)A_5 A_0)} \tag{40}$$

where is the modulus of the matrix, is the volume fraction of reinforcement, is the Poisson's ratio of the matrix, parameters, A_0, A_1, …, A_5 are functions of the Eshelby's tensor.

9.2.2.2 FINITE ELEMENT METHODS

The traditional framework in mechanics has always been the continuum. Under this framework, materials are assumed to be composed of a divisible continuous medium, with a constitutive relation that remains the same for a wide range of system sizes. Continuum equations are typically in the form of deterministic or stochastic partial differential equation (FDE's). The underling atomic structure of matter is neglected altogether and is replaced with a continuous and differentiable mass density. Similar replacement is made for other physical quantities such as energy and momentum. Differential equations are then formulated from basic physical principles, such as the conservation of energy or momentum. There are a large variety of numerical method that can be used for solving continuum partial differential equation, the most popular being the finite element method (FEM).

The finite element method is a numerical method for approximating a solution to a system of partial differential equations the FEM proceeds by dividing the continuum into a number of elements, each connected to the next by nodes. This discretization process converts the PDE's into a set of coupled ordinary equations that are solved at the nodes of the FE mesh and interpolated throughout the interior of the elements using shape functions. The main advantages of the FEM are its flexibility in geometry, refinement, and loading conditions. It should be noted that the FEM is local, which means that the energy within a body does not change throughout each element and only depends on the energy of the nodes of that element.

The total potential energy under the FE framework consists of two parts, the internal energy U and external work W.

$$\Pi = U - W \tag{41}$$

The internal energy is the strain energy caused by deformation of the body and can be written as:

$$U = \frac{1}{2}\int_{\Omega}\{\sigma\}^T\{\varepsilon\}d\Omega = \frac{1}{2}\int_{\Omega}\{\sigma\}^T[D]\{\varepsilon\}d\Omega \tag{42}$$

where $\{\sigma\} = \{\}^T$ denotes the stress vector; $\{\varepsilon\} = \{\}^T$ denotes the strain vector; $[D]$ is the elasticity matrix, and indicate that integration is over the entire domain.

The external work can be written as:

$$W = \int_{\Omega}\{uvw\}\begin{Bmatrix}\Im_x\\ \Im_y\\ \Im_z\end{Bmatrix}d\Omega + \int_{\Gamma}\{uvw\}\begin{Bmatrix}T_x\\ T_y\\ T_z\end{Bmatrix}d\Gamma \tag{43}$$

where u, v and w represent the displacement in x, y, z directions, respectively, is the force vector which contain both applied and body forces, is the surface traction vector, and indicates that the integration of the traction occurs only over the surface of the body. After discretization of the region into a number of elements, point – wise discretization of the displacements *u, v* and *z* directions, is achieved using shape function *[N]* for each element, such that the total potential energy becomes:

$$\Pi^g = \frac{1}{2}\{d\}^T \int_{\Omega^g}[B]^T [D][B]\{d\}d\Omega - \{d\}^T \int_{\Omega^g}[N]^T \begin{Bmatrix} \Im_x \\ \Im_y \\ \Im_z \end{Bmatrix} d\Omega - \{d\}^T \int_{\Gamma^g}[N]^T \begin{Bmatrix} T_x \\ T_y \\ T_z \end{Bmatrix} d\Gamma \quad (44)$$

where *[B]* is the matrix containing the derivation of the shape function, and *[d]* is a vector containing the displacements.

9.3 CARBON NANO TUBES (CNT): STRUCTURE AND PROPERTIES

A Carbon Nanotube is a tube shaped material, made of carbon, having length-to-diameter ratio over of 100,000,000:1, considerably higher than for any other material. These cylindrical carbon molecules have strange properties such as extraordinary mechanical, electrical properties and thermal conductivity, which are important for different fields of materials science and technology.

Nanotubes are members of the globular shape structural category. They are formed from rolling of one atom thick sheets of carbon called graphene, which cause to hollow structure. The process of rolling could be done at specific and discrete chiral angles, as nanotube properties are dependent to the rolling angle. The single nanotubes physically line up themselves into ropes situation together by Vander Waals forces.

Chemical bonding in nanotubes describes by orbital hybridization. The chemical bonding of nanotubes is constituted completely of sp^2 bonds, similar to those of graphite, which are stronger than the sp^3 bonds found in diamond, provide nanotubes with their unique strength.

9.3.1 CATEGORIZATION OF NANOTUBES

Carbon nanotube is classified as single walled nanotubes (SWNTs), multiwalled nanotubes (MWNTs) and Double walled carbon nanotubes (DWNTs).

9.3.1.1 SINGLE-WALLED CARBON NANOTUBE

SWNTs have a diameter of near to 1nanometerand tube length of longer than 10^9 times. The structure of a SWNT can be explained by wresting a graphene into a seamless cylinder. The way the graphene is wrested is depicted by a pair of indices *(n, m)*. The integer's *n* and *m* indicate the number of unit vectors in the direction of two points in the honey comb crystal lattice of graphene. If *n*=*m*, the nanotubes are called *armchair* nanotubes and while *m*= 0, the nanotubes are called *zigzag* nanotubes (Fig. 9.4). If not, they are named *chiral*. The *diameter* of a perfect nanotube can be calculated from its *(n, m)* indices as:

$$d = \frac{a}{\pi}\sqrt{(n^2 + nm + m^2)} = 78.3\sqrt{((n+m)^2 - nm}\,(pm)(a = 0.246\ \text{nm}) \quad (45)$$

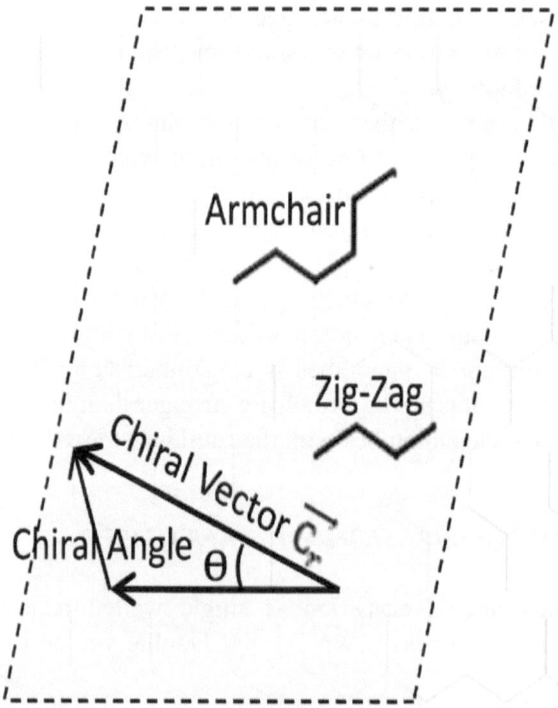

FIGURE 9.4 Schematic diagram of a hexagonal grapheme sheet.

SWNTs are an important kind of carbon nanotube due to most of their properties change considerably with the *(n, m)* values, and according to *Kataura plot*, this dependence is unsteady. Mechanical properties of single SWNTs were predicted remarkable by Quantum mechanics calculations as Young's modulus of 0.64–1 TPa, Tensile Strength of 150–180 GPa, strain to failure of 5–30% while having a relatively low density of 1.4–1.6 g/cm^3.

These high stiffness and superior mechanical properties for SWNTs are due to the chemical structure of the repeat unit. The repeat unit is composed completely of sp^{2}hybridized carbons and without any points for flexibility or rotation. Also, Single walled nanotubes with diameters of an order of a nanometer can be excellent conductors for electrical industries.

In most cases, SWNTs are synthesized through the reaction of a gaseous carbon feedstock to form the nanotubes on catalyst particles.

9.3.1.2 MULTI-WALLED CARBON NANOTUBE

Multi-walled nanotubes (MWNT) consist of several concentric graphene tubes, which its interlayer distance is around 3.4 Å; close to the distance between graphene layers in graphite. Its single shells in MWNTs can be explained as SWNTs.

9.3.1.3 DOUBLE-WALLED CARBON NANOTUBES (DWNT)

Double-walled carbon nanotubes consist of a particular set of nanotubes because their morphology and properties are comparable to those of SWNT but they have considerably superior chemical resistance. Prominently, it is important when grafting of chemical functions at the surface of the nanotubes is required to achieve new properties of the CNT. In this processing of SWNT, some *C=C* double bonds will be broken by covalent functionalization, and thus, both mechanical and electrical properties of nanotubes will be modified. About of DWNT, only the outer shell is modified.

9.3.2 CNT PROPERTIES

9.3.2.1 STRENGTH

In terms of strength, tensile strength and elastic modulus are explained and it has not been discovered any material as strong as carbon nanotubes yet. CNTs due to containing the single carbon atoms, having the covalent sp^2 bonds formed between them and they could resist against high tensile stress. Many studies have been done on tensile strength of carbon nanotubes and totally it was included that single CNT shells have strengths of about 100 GPa. Since density of a solid carbon nanotubes is around of 1.3 to 1.4 g/cm^3, specific strength of them is up to 48,000 kN·m·kg^{-1} which it causes to including carbon nanotube as the best of known materials, compared to high carbon steel that has specific strength of 154 kN·m·kg^{-1}.

Under extreme tensile strain, the tubes will endure plastic deformation, which means the deformation is invariable. This deformation commences at strains of around 5% and can enhance the maximum strain the tubes undergo before breakage by releasing strain energy.

Despite of the most high strength of single CNT shells, weak shear interactions between near shells and tubes leads to considerable diminutions in the effective strength of multiwalled carbon nanotubes, while crosslink in inner shells and tubes, included the strength of these materials is about 60 GPa for multiwalled carbon nanotubes and about 17 GPa for double-walled carbon nanotube bundles.

Almost, hollow structure and high aspect ratio of carbon nanotubes lead to their tendency to suffer bending when placed under compressive, torsion stress.

9.3.2.2 HARDNESS

The regular single-walled carbon nanotubes have ability to undergo a transformation to great hard phase and therefore, they can endure a pressure up to 25 GPa without deformation. The bulk modulus of great hard phase nanotubes is around 500 GPa, which is higher than that of diamond (420 GPa for single diamond crystal).

9.3.2.3 ELECTRICAL PROPERTIES

Because of the regularity and exceptional electronic structure of graphene, the structure of a nanotube affects its electrical properties strongly. It has been concluded that for a given *(n, m)* nanotube, while $n=m$, the nanotube is metallic; if $n-m$ is a multiple of 3, then the nanotube is semiconducting with a very small band gap, if not, the nanotube is a moderate semiconductor. However, some exceptions are in this rule, because electrical properties can be strongly affected by curvature in small diameter carbon nanotubes. In theory, metallic nanotubes can transmit an electric current density of 4 × 109A/cm^2, which is more than 1,000 times larger than those of metals such as copper, while electro migration lead to limitation of current densities for copper interconnects.

Because of nanoscale cross-section in carbon nanotubes, electrons spread only along the tube's axis and electron transfer includes quantum effects. As a result, carbon nanotubes are commonly referred to as one-dimensional conductors. The maximum electrical transmission of a SWNT is $2G_0$, where $G_0= 2e^2/h$ is the transmission of a single ballistic quantum channel.

9.3.2.4 THERMAL PROPERTIES

It is expected that nanotubes act as very good thermal conductors, but they are good insulators laterally to the tube axis. Measurements indicate that SWNTs have a room temperature thermal conductivity along its axis of about 3500 $W{\cdot}m^{-1}{\cdot}K^{-1}$; higher than that for copper (385 $W{\cdot}m^{-1}{\cdot}K^{-1}$). Also, SWNT has a room temperature thermal conductivity across its axis (in the radial direction) of around 1.52 $W{\cdot}m^{-1}{\cdot}K^{-1}$, which is nearly as thermally conductive as soil. The temperature constancy of carbon nanotubes is expected to be up to 2800 °C in vacuum and about 750 °C in air.

9.3.2.5 DEFECTS

As with any material, the essence of a crystallographic defect affects the material properties. Defects can happen in the form of atomic vacancies. High levels of such defects can drop the tensile strength up to 85%. A main example is the Stone Wales defect, which makes a pentagon and heptagon pair by reorganization of the bonds. Having small structure in carbon nanotubes lead to dependency of their tensile strength to the weakest segment.

Also, electrical properties of CNTs can be affected by crystallographic defects. A common result is dropped conductivity through the defective section of the tube. A defect in conductive nanotubes can cause the adjacent section to become semiconducting, and particular monatomic vacancies induce magnetic properties.

Crystallographic defects intensively affect the tube's thermal properties. Such defects cause to phonon scattering, which in turn enhance the relaxation rate of the phonons. This decreases the mean free path and declines the thermal conductivity of nanotube structures. Phonon transport simulations show that alternative defects such as nitrogen or boron will mainly cause to scattering of high frequency optical phonons. However, larger scale defects such as Stone Wales defects lead to phonon scattering over a wide range of frequencies, causing to a greater diminution in thermal conductivity.

9.3.3 METHODS OF CNT PRODUCTION

9.3.3.1 ARC DISCHARGE METHOD

Nanotubes were perceived in 1991 in the carbon soot of graphite electrodes during an arc discharge, by using a current of 100amps that was intended to create fullerenes. However, for the first time, macroscopic production of carbon nanotubes was done in 1992 by the similar method of 1991. During this process, the carbon included the negative electrode sublimates due to the high discharge temperatures. As the nanotubes were initially discovered using this technique, it has been the most widely used method for synthesis of CNTs. The revenue for this method is up to 30% by weight and it produces both single and multi walled nanotubes with lengths of up to 50 micrometers with few structural defects.

9.3.3.2 LASER ABLATION METHOD

Laser ablation method was developed by Dr. Richard Smalley and coworkers at Rice University. In that time, they were blasting metals with a laser to produce a variety of metal molecules. When they noticed the existence of nanotubes they substituted the metals with graphite to produce multiwalled carbon nanotubes. In the next year, the team applied a composite of graphite and metal catalyst particles to synthesize single walled carbon nanotubes. In laser ablation method, vaporizing a graphite target in a high temperature reactor is done by a pulsed laser while an inert gas is bled into the chamber. Nanotubes expand on the cooler surfaces of the reactor as the vaporized carbon condenses. A water-cooled surface may be contained in the system to gathering the nanotubes.

The laser ablation method revenues around 70% and produces principally single-walled carbon nanotubes with a controllable diameter determined by the reaction temperature. However, it is more costly than either arc discharge or chemical vapor deposition.

9.3.3.3 PLASMA TORCH

In 2005, a research group from the University of Sherbrook and the National Research Council of Canada could synthesize Single walled carbon nanotubes by the induction thermal plasma method. This method is alike to arc discharge in that both apply ionized gas to achieve the high temperature necessary to vaporize carbon containing substances and the metal catalysts necessary for the following nanotube development. The thermal plasma is induced by high frequency fluctuating currents in a coil, and is kept in flowing inert gas. Usually, a feedstock of carbon black and metal catalyst particles is supplied into the plasma, and then cooled down to constitute single walled carbon nanotubes. Various single wall carbon nanotube diameter distributions can be synthesized.

The induction thermal plasma method can create up to 2 g of nanotube material per minute, which is higher than the arc-discharge or the laser ablation methods.

9.3.3.4 CHEMICAL VAPOR DEPOSITION (CVD)

In 1952 and 1959, the catalytic vapor phase deposition of carbon was studied, and finally, in1993; the carbon nanotubes were constituted by this process. In 2007, researchers at the University of Cincinnati (UC) developed a process to develop aligned carbon nanotube arrays of 18 mm length on a First Nano ET3000 carbon nanotube growth system.

In CVD method, a substrate is prepared with a layer of metal catalyst particles, most usually iron, cobalt, nickel or a combination. The metal nanoparticles can also be formed by other ways, including reduction of oxides or oxides solid solutions. The diameters of the carbon nanotubes which are to be grown are related to the size of the metal particles. This can be restrained by patterned (or masked) deposition of the metal, annealing, or by plasma etching of a metal layer. The substrate is heated to around of 700 °C. To begin the enlargement of nanotubes, two types of gas are bled into the reactor: a process gas (such as ammonia, nitrogen or hydrogen) and a gas containing carbon (such as acetylene, ethylene, ethanol or methane). Nanotubes develop at the sites of the metal catalyst; the gas containing carbon is broken apart at the surface of the catalyst particle, and the carbon is transferred to the edges of the particle, where it forms the nanotubes. The catalyst particles can stay at the tips of the growing nanotube during expansion, or remain at the nanotube base, depending on the adhesion between the catalyst particle and the substrate.

CVD is a general method for the commercial production of carbon nanotubes. For this idea, the metal nanoparticles are mixed with a catalyst support such as MgO or Al_2O_3 to enhance the surface area for higher revenue of the catalytic reaction of the carbon feedstock with the metal particles. One matter in this synthesis method is the removal of the catalyst support via an acid treatment, which sometimes could destroy the primary structure of the carbon nanotubes. However, other catalyst supports that are soluble in water have verified effective for nanotube development.

Of the different means for nanotube synthesis, CVD indicates the most promise for industrial scale deposition, due to its price/unit ratio, and because CVD is capable of increasing nanotubes directly on a desired substrate, whereas the nanotubes must be collected in the other expansion techniques. The development sites are manageable by careful deposition of the catalyst.

9.3.3.5 SUPER-GROWTH CVD

Researchers developed super growth CVD (water assisted chemical vapor deposition), by adding water into CVD reactor to improve the activity and lifetime of the catalyst. Dense millimeter tall nanotube *forests*, aligned normal to the substrate, were created. The *forests* expansion rate could be extracted, as:

$$H(t) = \beta\tau_0(1 - e^{-\frac{t}{\tau_0}}) \tag{46}$$

where, β is the initial expansion rate andis the characteristic catalyst lifetime.

9.4 SIMULATION OF CNT'S MECHANICAL PROPERTIES

9.4.1 *MODELING TECHNIQUES*

The theoretical efforts in simulation of CNT mechanical properties can be categorized in three groups as follow:

- atomistic simulation;
- continuum simulation;
- nano-scale continuum modeling.

9.4.1.1 ATOMISTIC SIMULATION

Based on interactive forces and boundary conditions, atomistic modeling predicts the positions of atoms. Atomistic modeling techniques can be classified into three main categories, namely the MD, MC and Ab initio approaches. Other atomistic modeling techniques such as TBMD, LD, DFT, Morse potential function model, and modified Morse potential function model were also applied as discussed in last section.

The first technique used for simulating the behaviors of CNTs was MD method. This method uses realistic force fields (many-body interatomic potential functions) to determination the total energy of a system of particles. Whit the calculation of the total potential energy and force fields of a system, the realistic calculations of the behavior and the properties of a system of atoms and molecules can be acquired. Although the main aspect of both MD and MC simulations methods is based on second Newton's law, MD methods are deterministic approaches, in comparison to the MC methods that are stochastic ones.

In spite the MD and MC methods depend on the potentials that the forces acting on atoms by differentiating inter atomic potential functions, the Ab initio techniques are accurate methods which are based on an accurate solution of the Schrödinger equation. Furthermore, the Ab initio techniques are potential-free methods wherein the atoms forces are determined by electronic structure calculations progressively.

In generally, MD simulations provide good predictions of the mechanical properties of CNTs under external forces. However, MD simulations take long times to produce the results and consumes a large amount of computational resources, especially when dealing with long and multiwalled CNTs incorporating a large number of atoms.

9.4.1.2 CONTINUUM MODELING

Continuum mechanics-based models are used by many researches to investigate properties of CNTs. The basic assumption in these theories is that a CNT can be modeled as a continuum structure which has continuous distributions of mass, density, stiffness, etc. Therefore, the lattice structure of CNT is simply neglected in and it is replaced with a continuum medium. It is important to meticulously investigate the validity of continuum mechanics approaches for modeling CNTs, which the real discrete nano-structure of CNT is replaced with a continuum one. The continuum modeling can be either accomplished analytical (micromechanics) or numerically representing FEM.

Continuum shell models used to study the CNT properties and showed similarities between MD simulations of macroscopic shell model. Because of the neglecting the discrete nature of the CNT geometry in this method, it has shown that mechanical properties of CNTs were strongly dependent on atomic structure of the tubes and like the curvature and chirality effects, the mechanical behavior of CNTs cannot be calculated in an isotropic shell model. Different from common *shell model,* which is constructed as an isotropic continuum shell with constant elastic properties for SWCNTs, the MBASM model can predict the chirality induced anisotropic effects on some mechanical behaviors of CNTs by incorporating molecular and continuum mechanics solutions. One of the other theory is shallow shell theories, this theory are not accurate for CNT analysis because of CNT is a nonshallow structure. Only more complex shell is capable of reproducing the results of MD simulations.

Some parameters, such as wall thickness of CNTs are not well defined in the continuum mechanics. For instance, value of 0.34 nm, which is interplanar spacing between graphene sheets in graphite, is widely used for tube thickness in many continuum models.

The finite element method works as the numerical methods for determining the energy minimizing displacement fields, while atomistic analysis is used to determine the energy of a given configuration. This is in contrast to normal finite element approaches, where the constitutive input is made via phenomenological models. The method is successful in capturing the structure and energetic of dislocations. Finite element modeling is directed by using 3D beam element, which is as equivalent beam to construct the CNT. The obtained results will be useful in realizing interactions between the nanostructures and substrates and also designing composites systems.

9.4.1.3 NANO-SCALE CONTINUUM MODELING

Unlike to continuum modeling of CNTs where the entirely discrete structure of CNT is replaced with a continuum medium, nano-scale continuum modeling provides a rationally acceptable compromise in the modeling process by replacing C–C bond with a continuum element. In the other hand, in nano-scale continuum modeling the molecular interactions between C–C bonds are captured using structural members whose properties are obtained by atomistic modeling. Development of nano-scale continuum theories has stimulated more excitement by incorporating continuum mechanics theories at the scale of nano. Nano-scale continuum modeling is usually accomplished numerically in the form of finite element modeling. Different elements consisting of rod, truss, spring and beam are used to simulate C–C bonds. The two common method of nano-scale continuum are *quasi-continuum* (QC) and *equivalent-continuum* methods, which have been used in nano-scale continuum modeling.

The QC method which presents a relationship between the deformations of a continuum with that of its crystal lattice uses the classical Cauchy–Born rule and representative atoms. The quasi-continuum method mixes atomistic-continuum formulation and is based on a finite element discretization of a continuum mechanics variation principle.

The equivalent continuum method developed by providing a correlation between computational chemistry and continuum structural mechanics. It has considered being equal total molecular potential energy of a nanostructure with the strain energy of its equivalent continuum elements.

This method has been proposed for developing structure-property relationships of nano-structured materials and works as a link between computational chemistry and solid mechanics by substituting discrete molecular structures with equivalent-continuum models. It has been shown that this substitution may be accomplished by equating the molecular potential energy of a nano-structured material with the strain energy of representative truss and continuum models.

Because of the approach uses the energy terms that are associated with molecular mechanics modeling, a brief description of molecular mechanics is given first followed by an outline of the equivalent-truss and equivalent-continuum model development.

9.4.1.3.1 MOLECULAR MECHANICS

An important part in molecular mechanics calculations of the nano-structure materials is the determination of the forces between individual atoms. This description is characterized by a force field. In the most general form, the total molecular potential energy, *E*, for a nano-structured material is described by the sum of many individual energy contributions:

$$E_{total} = \sum E_\rho + \sum E_\theta + \sum E_\tau + \sum E_\omega + \sum E_{vdw} + \sum E_{el} \tag{47}$$

Where , , , are the energies associated with bond stretching, angle variation, torsion, and inversion, respectively. The atomic deformation mechanisms are illustrated in Figs. 9.5 and 9.6.

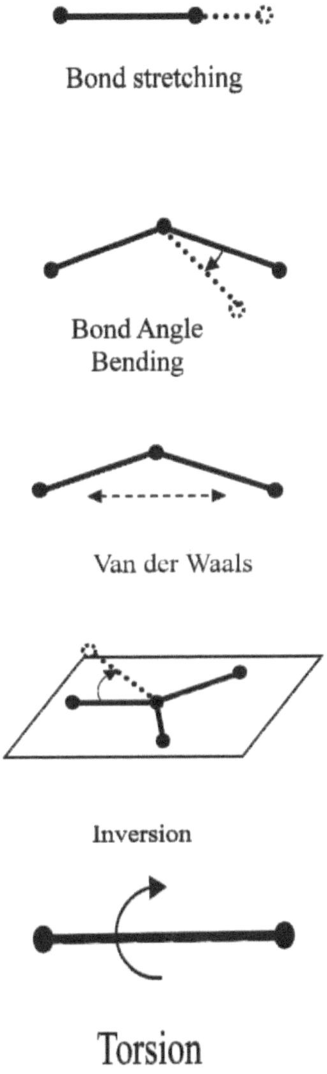

FIGURE 9.5 Atomistic bond interaction mechanisms.

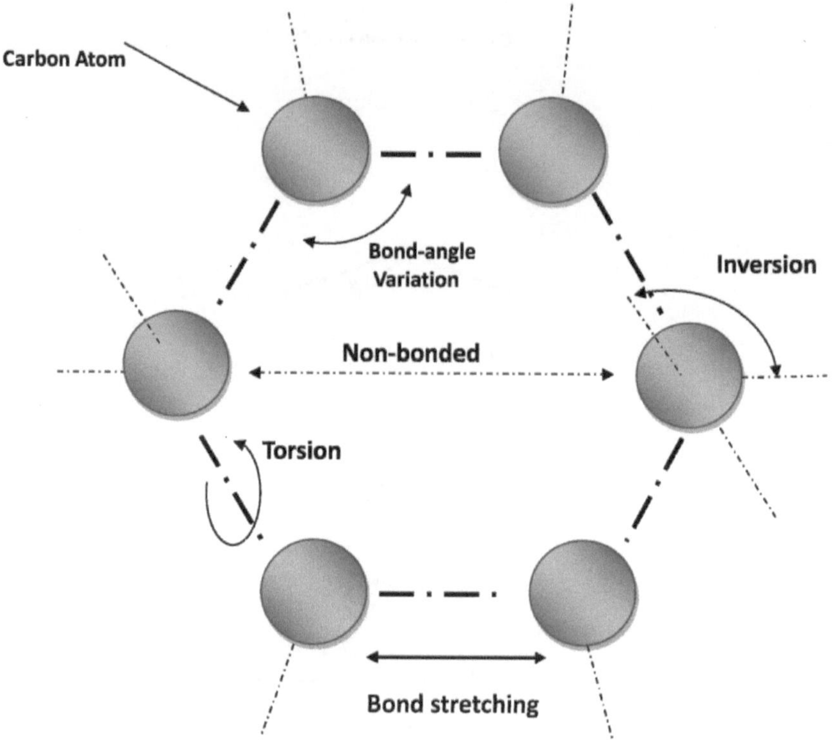

FIGURE 9.6 Schematic interaction for carbon atoms.

The nonbonded interaction energies consist of Vander Waalsand electrostatic, terms. Depending on the type of material and loading conditions various functional forms may be used for these energy terms. In condition where experimental data are either unavailable or very difficult to measure, quantum mechanical calculations can be a source of information for defining the force field.

In order to simplify the calculation of the total molecular potential energy of complex molecular structures and loading conditions, the molecular model substituted by intermediate model with a pin-jointed Truss model based on the nature of molecular force fields, to represent the energies given by Eq. (4.1), where each truss member represents the forces between two atoms.

So, a truss model allows the mechanical behavior of the nano-structured system to be accurately modeled in terms of displacements of the atoms. This mechanical representation of the lattice behavior serves as an intermediate step in linking the molecular potential with an equivalent-continuum model.

In the truss model, each truss element represented a chemical bond or a nonbonded interaction. The stretching potential of each bond corresponds with the stretching of the corresponding truss element. Atoms in a lattice have been viewed as masses that are held in place with atomic forces that is similar to elastic springs. Therefore, bending of Truss elements is not needed to simulate the chemical bonds, and it is assumed that each truss joint is pinned, not fixed, Fig. 9.7 shown the atomistic-based continuum modeling and RVE of the chemical, truss and continuum models.

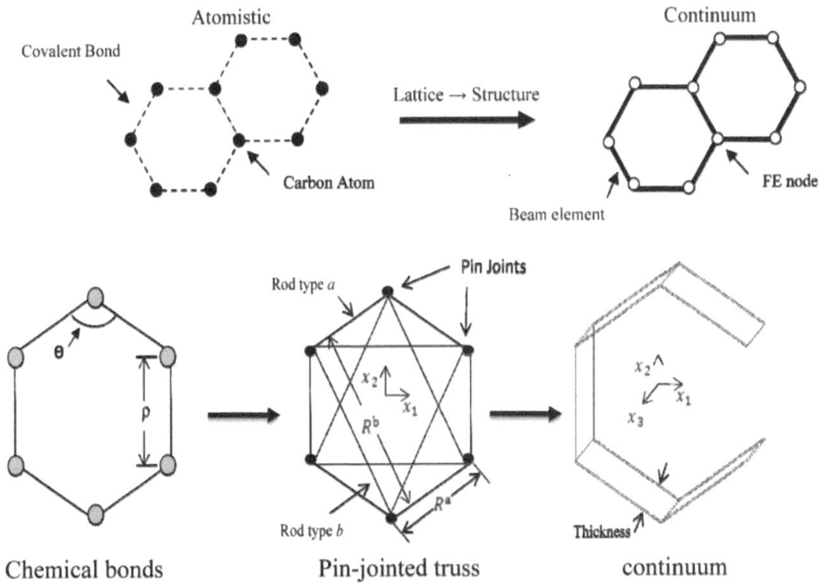

FIGURE 9.7 Atomistic-based continuum modeling and RVE of the chemical, truss and continuum models.

The mechanical strain energy, , of the truss model is expressed in the form:

$$\Lambda^t = \sum_n \sum_m \frac{A_m^n Y_m^n}{2R_m^n} (r_m^n - R_m^n)^2 \tag{48}$$

where, and are the cross-sectional area and Young's modulus, respectively, of rod *m* of truss member type *n*. The term is the stretching of rod *m* of truss member type n, where and are the undeformed and deformed lengths of the truss elements, respectively.

In order to represent the chemical behavior with the truss model, Eq. (47) must be consider being equal with Eq. (48) in a physically meaningful manner. Each of the two equations is a sum of energies for particular degrees of freedom. In comparison of the Eq. (47) that have bond angle variance and torsion terms the Eq. (48) has stretching term only, so it made the main difficulty for substitute these tow equations. No generalization can be made for overcoming this difficulty for every nano-structured system. It means that possible solution must be determined for a specific nano-Structured material depending on the geometry, loading conditions, and degree of accuracy sought in the model.

9.5 LITERATURE REVIEW ON CNT SIMULATION

Different researchers have been doing many efforts to simulate mechanical properties of CNT, in generally the main trends of these methods employed by different researchers to predict the elastic modulus of SWCNTs results in terms of three main parameters of morphology: radius, chirality, and wall thickness. The dependency of results to the diameter of CNT becomes less pronounced when non-linear inter atomic potentials are employed instead of linear ones.

There are plenty of experimental and theoretical techniques for characterizing Young's modulus of carbon nanotubes, all three main group of CNT modeling: MD, CM and NCM have been used in the literature. Some of recent different theoretical methods for predicting Young's modulus of carbon nanotubes are summarized in Tables 9.1–9.3.

TABLE 9.1 MD Methods for Prediction Young' Modulus of CNT

Researchers	Year	Method	Young's modulus (TPa)	Results
Liew et al. [1]	2004	Second generation reactive of empirical bond-order (REBO)	1.043	Examining the elastic and plastic properties of CNTs under axial tension
H. W. Zhang [2]	2005	Modified Morse Potentials and Tersoff–Brenner potential	1.08	Predicting the elastic properties of SWCNTs based on the classical Cauchy–Born rule
Agrawal et al. [3]	2006	A combination of a second generation reactive empirical bond order potential and vdW interactions	0.55, 0.73, 0.74, 0.76	Predicting Young's modulus by four MD approaches for an armchair (14,14) type SWCNT and investigating effect of defects in the form of vacancies, van der Waals (vdW) interactions, chirality, and diameter
Cheng et al. [4]	2008	MD simulations using Tersoff– Brenner potential to simulate covalent bonds while using Lennard–Jones to model interlayer interactions	1.4 for armchair 1.2 for zigzag	Evaluating the influence of surface effect resulting in relaxed unstrained deformation and in-layer nonbonded interactions using atomistic continuum modeling approach
Cai et al. [5]	2009	Linear scaling self-consistent charge, density functional tight binding (SCC-DFTB) and an Ab initio Dmol3	—	Investigating the energy and Young's modulus as a function of tube length for (10, 0) SWCNTs
Ranjbartoreh and Wang[6]	2010	Large-scale atomic/molecular massively parallel simulator (LAMMPS) code	0.788 for armchair 1.176 for zigzag	Effects of chirality and Van der Waals interaction on Young's modulus, elastic compressive modulus, bending, tensile, compressive stiffness, and critical axial force of DWCNTs

TABLE 9.2 Continuum Simulation Methods for Prediction Young' Modulus of CNT

Researchers	Year	Method	Young's modulus (TPa)	Results
Sears and Batra [7]	2004	Equivalent continuum tube	2.52	Results of the molecular-mechanics simulations of a SWNT is used to derive the thickness and values of the two elastic modulus of an isotropic linear elastic cylindrical tube equivalent to the SWNT
Wang [8]	2004	Stretching and rotating springs, equivalent continuum plate	0.2–0.8 for zigzags 0.57–0.54 for armchairs	Effective in-plane stiffness and bending rigidity of SWCNTs
Kalamkarov et al. [9]	2006	Asymptotic homogenization, cylindrical network shell	1.71	Young's modulus of SWCNTs
Gupta and Batra [10]	2008	Equivalent continuum structure	0.964 ± 0.035	Predicting Young's modulus of SWCNTs b equating frequencies of axial and torsional modes of vibration of the ECS to those of SWCNT computed through numerical simulations using the MM3 potential
Giannopoulos et al. [11]	2008	FEM, spring elements	1.2478	Young's and Shear modulus of SWCNTs using three dimensional spring-like elements
Papanikos et al. [12]–	2008	3-D beam element	0.4–2.08	3D FE analysis, assuming a linear behavior of the C–C bonds

TABLE 9.3 Nano-scale Continuum Methods for Prediction Young' Modulus of CNT

Researchers	Year	Method	Young's modulus (TPa)	Results
Li and Chou [13]	2003	Computational model using beam element and nonlinear truss rod element	1.05 ± 0.05	Studying of elastic behavior of MWCNTs and investigating the influence of diameter, chirality and the number of tube layers on elastic modulus
Natsuki et al. [14]	2004	Frame structure model with a spring constant for axial deformation and bending of C–C bond using analytical method	0.61–0.48	A two dimensional continuum-shell model which is composed of the discrete molecular structures linked by the carbon–carbon bonds
Natsuki and Endo [15]	2004	Analytical analysis by exerting nonlinear Morse potential	0.94 for armchair (10,10)	Mechanical properties of armchair and zigzag CNT are investigated. The results show the atomic structure of CNT has a remarkable effect on stress– strain behavior
Xiao et al. [16]	2005	Analytical molecular structural mechanics based on the Morse potential	1–1.2	Young's modulus of SWCNTs under tension and torsion and results are sensitive to the tube diameter and the helicity
Tserpes and Papanikos [17]	2005	FE method employing 3D elastic beam element	For chirality (8,8) and thickness 0.147 nm 2.377	Evaluating Young's modulus of SWCNT using FEM and investigating the influence of wall thickness, tube diameter and chirality on CNT Young's modulus

TABLE 9.3 *(Continued)*

Jalalahmadi and Naghdabadi [18]	2007	Finite element modeling employing beam element based on Morse potential	3.296, 3.312, 3.514	Predicting Young's modulus utilizing FEM and Morse potential to obtain mechanical properties of beam elements, Moreover investigating of wall thickness, diameter and chirality effects on Young's modulus of SWCNT
Meo and Rossi [19]	2007	Nonlinear and torsion springs	0.912 for zigzag structures 0.920 for armchair structures	Predicting the ultimate strength and strain of SWCNTs and effects of chirality and defections
PourAkbar Saffar et al. [20]	2008	Finite element method utilizing 3D beam element	1.01 for (10,10)	Obtaining Young modulus of CNT to use it in FE analysis and investigating Young modulus of CNT reinforced composite
Cheng et al. [21]	2008	MD simulations using Tersoff– Brenner potential to simulate covalent bonds while using Lennard–Jones to model interlayer interactions. Finite element analysis employing nonlinear spring element and equivalent beam element to model in-layer non-bonded interactions and covalent bond of two neighbor atoms, respectively	1.4 for armchair and 1.2 for zigzag	Evaluating the influence of surface effect resulting in relaxed unstrained deformation and in-layer nonbonded interactions using atomistic continuum modeling approach

TABLE 9.3 *(Continued)*

Avila and Lacerda [22]	2008	FEM using 3D beam element to simulate C–C bond	0.95–5.5 by altering the CNT radius Constructing SWCNT (armchair, zigzag and chiral)	Model based on molecular mechanic and evaluating its Young's modulus and Poisson's ratio
Wernik and Meguid [23]	2010	FEM using beam element to model the stretching component of potential and also nonlinear rotational spring to take account the angle-bending component	0.9448 for armchair (9, 9)	Investigating nonlinear response of CNT using modified Morse potential, also studying the fracture process under tensile loading and torsional bucking
Shokrieh and Rafiee [24]	2010	Linking between inter-atomic potential energies of lattice molecular structure and strain energies of equivalent discrete frame structure	1.033–1.042	A closed-form solution for prediction of Young's modulus of a graphene sheet and CNTs and finite element modeling of CNT using beam element
Lu and Hu [25]	2012	Using nonlinear potential to simulate C–C bond with considering elliptical-like cross section area of C–C bond	For diameter range 0.375–1.8, obtained Young's modulus is 0.989–1.058	Predicting mechanical properties of CNT using FEM, also investigating rolling energy per atom to roll graphene sheet to SWNT
Rafiee and Heidarhaei [26]	2012	Nonlinear FEM using nonlinear springs for both bond stretching and bond angle variations	1.325	Young's modulus of SWCNTs and investigating effects of chirality and diameter on that

9.5.1 BUCKLING BEHAVIOR OF CNTS

One of the technical applications on CNT related to buckling properties is the ability of nanotubes to recover from elastic buckling, which allows them to be used several times without damage. One of the most effective parameters on buckling behaviors of CNT on compression and torsion is chirality that thoroughly investigated by many researchers [27, 28]. Chang et al. [29] showed that zigzag chirality is more stable than armchair one with the same diameter under axial compression. Wang et al. [30] reviewed buckling behavior of CNTs recently based on special characteristics of buckling behavior of CNTs.

The MD simulations have been used widely for modeling buckling behavior of CNTs [31–35]. MD use simulation package which included Newtonian equations of motion based on Tersoff–Brenner potential of inter atomic forces. Yakobson et al. [36] used MD simulations for buckling of SWCNTs and showed that CNTs can provide extreme strain without permanent deformation or atomic rearrangement. Molecular dynamics simulation by destabilizing load that composed of axial compression, torsion have been used to buckling and postbucking analysis of SWCNT [37].

The buckling properties and corresponding mode shapes were studied under different rotational and axial displacement rates which indicated the strongly dependency of critical loads and buckling deformations to these displacement rates. Also buckling responses of multiwalled carbon nanotubes associated to torsion springs in electromechanical devices were obtained by Jeong [38] using classical MD simulations.

Shell theories and molecular structural simulations have been used by some researchers to study buckling behavior and CNT change ability for compression and torsion. Silvestre et al. [39] showed the inability of Donnell shell theory [40] and shown that the Sanders shell theory [41] is accurate in reproducing buckling strains and mode shapes of axially compressed CNTs with small aspect ratios. It's pointed out that the main reason for the incorrectness of Donnell shell model is the inadequate kinematic hypotheses underlying it.

It is exhibited that using Donnell shell and uniform helix deflected shape of CNT simultaneously, leads to incorrect value of the critical angle of twist, conversely Sanders shell model with non uniform helix deflected shape presents correct results of critical twist angle. Besides, Silvestre et al. [42] presented an investigation on linear buckling and postbuckling behavior of CNTs using molecular dynamic simulation under pure shortening and twisting.

Ghavamian and Öchsner [43] were analyzed the effect of defects on the buckling behavior of single and multiwalled CNT based on FE method considering three most likely atomic defects including impurities, vacancies (carbon vacancy) and introduced disturbance. The results demonstrate that the existence of any type kinks in CNTs structure, conducts to lower critical load and lower buckling properties.

Zhang et al. [44] performed an effort on the accuracy of the Euler Bernoulli beam model and Donnell shell model and their nonlocal counterparts in predicting the buckling strains of single-walled CNTs. Comparing with MD simulation results, they concluded that the simple Euler–Bernoulli beam model is sufficient for predicting the buckling strains of CNTs with large aspect ratios (i.e., length to diameter ratio $L/d > 10$). The refined Timoshenko's beam model for nonlocal beam theory is needed for CNTs with intermediate aspect ratios (i.e., $8 < L/d < 10$). The Donnell thin shell theory is unable to capture the length dependent critical strains obtained by MD simulations for CNTs with small aspect ratios (i.e., $L/d < 8$) and hence this simple shell theory is unable to model small aspect ratio CNTs. Tables 9.4–9.6 summarizes some of theoretical simulations of buckling behavior of CNTs.

9.5.2 VIBRATIONS ANALYSIS

Dynamic mechanical behaviors of CNTs are of importance in various applications, such as high frequency oscillators and sensors [56]. By adding CNTs to polymer the fundamental frequencies of CNT reinforced polymer can be improved remarkably without significant change in the mass density of material [57]. It is importance indicate that the dynamic mechanical analysis confirms strong influence of CNTs on the composite damping properties [58].

The simulation methods of CNT's vibrating were reviewed by Gibson et al. [59] in 2007. Considering wide applications of CNTs, receiving natural frequencies and mode shapes by assembling accurate theoretical model. For instance, the oscillation frequency is a key property of the resonator when CNTs are used as nano mechanical resonators. Moreover, by

TABLE 9.4 MD Simulation of CNT Buckling

Researchers	Year	Method	Chairality	Length (nm)	Diameter (nm)	Results
Wang et al. [30]	2005	Tersoff–Brenner potential	(n,n)(n,0)	7–19	0.5–1.7	Obtaining critical stresses and comparing between the buckling behavior in nano and macroscopic scale
Sears and Batra [45]	2006	MM3 class II pair wise potential, FEM, equivalent continuum structures using Euler buckling theory	Various zigzag CNTs	50 - 350Å	—	Buckling of axially compressed multiwalled carbon nanotubes by using molecular mechanics simulations and developing continuum structures equivalent to the nanotubes
Xin et al. [46]	2007	Molecular dynamics simulation using Morse potential, harmonic angle potential and Lennard–Jones potential	(7,7) and (12,0)	—	—	Studying buckling behavior of SWCNTs under axial compression based on MD method, also evaluating the impression of tube length, temperature and configuration of initial defects on mechanical properties of SWCNTs
Silvestre et al. [39]	2011	The second generation reactive empirical bond order (REBO) potential	(5, 5)(7,7)	—	—	Comparison between buckling behavior of SWCNTs with Donnell and Sanders shell theories and MD results
Ansari et al. [35]	2011	"AIREBO" potential	(8,8)(14, 0)	—	—	Axial buckling response of SWCNTs based on nonlocal elasticity continuum with different BCs and extracting nonlocal
Silvestre et al. [47]	2012	Tresoff-Bernner Covalent Potential	(8,8) (5,5) (6,3)	—	—	Buckling behavior of the CNTs under Pure shortening and Pure twisting as well as their pre-critical and post-critical stiffness

TABLE 9.5 Continuum Simulation of CNT Buckling

Researchers	Year	Method	Chairality	Length (nm)	Diameter (nm)	Results
He et al. [48]	2005	Continuum cylindrical shell	—			Establishing an algorithm for buckling analysis of multi-walled CNTs based on derived formula which considering Van der Waals interaction between any two layers of MWCNT
Ghorbanpour Arani et al. [49]	2007	FEM, cylindrical shell	—	11.2 and 44.8	3.2	Pure axially compressed buckling and combined loading effects on SWCNT
Yao et al. [50]	2008	FEM, elastic shell theory	—	—	0.5–3	Bending buckling of single- double- and multi walled CNTs
Guo et al. [51]	2008	Atomic scale finite element method	(15,0) (10,0)	8.3 and 8.38	—	Bending buckling of SWCNTs
Chan et al [52]	2011	Utilizing Donnell shell equilibrium equation and also Euler–Bernoulli beam equation incorporating curvature effect	Various chairality	—	—	Investigating pre and post-buckling behavior of MWCNTs and multi-walled carbon nano peapods considering vdW interactions between the adjacent walls of the CNTs and the interactions between the fullerenes and the inner wall of the nanotube
Silvestre [53]	2012	Donnell and Sanders shell model with non-uniform helix deflected shape		Various length to diameter ratio	Various length to diameter ratio	Investigating of critical twist angle of SWCNT and comparing the accuracy of two Donnell and Sanders shell model with uniform and non-uniform helix deflected shape

TABLE 9.6 Nano-scale Continuum Simulation of CNT Buckling

Researchers	Year	Method	Chairality	Length (nm)	Diameter (nm)	Results
Li and Chou [54]	2004	Molecular structural mechanics with 3D space frame-like structures with beams	(3,3)(8,8) (5,0)(14,0)	—	0.4–1.2	Buckling behaviors under either compression or bending for both single-and double-walled CNTs
Chang et al. [29]	2005	Analytical molecular mechanics, potential energy	—	—	1–5	Critical buckling strain under axial compression
Hu et al. [55]	2007	Molecular structural mechanics with 3D beam elements	Various type of armchair and zigzag	Various length to diameter ratio	Various length to diameter ratio	Investigating of buckling characteristics of SWCNT and DWCNT by using the beam element to model C–C bond and proposed rod element to model vdW forces in MWCNT, also the validity of Euler's beam buckling theory and shell buckling mode are studied

using accurate theoretical model to acquire natural frequencies and mode shapes, the elastic modulus of CNTs can be calculated indirectly.

Timoshenko's beam model used by Wang et al. [60] for free vibrations of MWCNTs study; it was shown that the frequencies are significantly over predicted by the Euler's beam theory when the aspect ratios are small and when considering high vibration modes. They indicated that the Timoshenko's beam model should be used for a better prediction of the frequencies especially when small aspect ratio and high vibration modes are considered.

Hu et al. [55] presented a review of recent studies on continuum models and MD simulations of CNT's vibrations briefly [61]. Three constructed model of SWCNT consisting of Timoshenko's beam, Euler–Bernoulli's beam and MD simulations are investigated and results show that fundamental frequency decreases as the length of a SWCNT increases and also the Timoshenko's beam model provides a better prediction of short CNT's frequencies than that of Euler–Bernoulli beam's model. Comparing the fundamental frequency results of transverse vibrations of cantilevered SWCNTs it can be seen that both beam models are not able to predict the fundamental frequency of cantilevered SWCNTs shorter than 3.5 nm.

An atomistic modeling technique and molecular structural mechanics are used by Li and Chou [62] to calculated fundamental natural frequency of SWCNTs. A free and forced vibrations of SWCNT have been assessed by Arghavan and Singh [63] using space frame elements with extensional, bending and torsional stiffness properties to modeling SWCNTs and compared their results with reported results by of Sakhaee pour [64] and Li and Chou [62]. Their results were in close agreement with two other results, within three to five percent.

Furthermore, free vibrations of SWCNTs have been studied by Gupta et al. [65] using the MM3 potential. They mentioned that calculations based on modeling SWCNTs as a beam will overestimate fundamental frequencies of the SWCNT. They derived the thickness of a SWCNT/shell to compare the frequencies of a SWCNT obtained by MM simulations with that of a shell model and provided an expression for the wall thickness in terms of the tube radius and the bond length in the initial relaxed configuration of a SWCNT. Aydogdu [66] showed that axial vibration frequencies of SWCNT embedded in an elastic medium highly over estimated by the classical rod model because of ignoring the effect of small length scale and developed an elastic rod model based on local and nonlocal rod theories to investigate the small scale effect on the axial vibrations of SWCNTs.

The vibration properties of two and three functioned of carbon nanotubes considering different boundary conditions and geometries were studied by Fakhrabadi et al. [67]. The results show that the tighter boundaries, larger diameters and shorter lengths lead to higher natural frequencies. Lee and Lee [68] fulfilled modal analysis of SWCNT's and nano cones (SWCNCs) using finite element method with ANSYS commercial package. The vibration behaviors of SWCNT with fixed beam and cantilever boundary conditions with different cross-section types consisting of circle and ellipse were constructed using 3D elastic beams and point masses. The nonlinear vibration of an embedded CNT's were studied by Fu et al.[69] and the results reveal that the nonlinear free vibration of nano-tubes is effected significantly by surrounding elastic medium.

The same investigation has been accomplished by Ansari and Hemmatnezhad [70] using the variational iteration method (VIM). Wang et al. [71]have studied axis symmetric vibrations of SWCNT immerged in water which in contrast to solid liquid system, a submerged SWCNT is coupled with surrounding water via vdW interaction. The analysis of DWCNTs vibration characteristics considering simply support boundary condition are carried out by Natsuki et al. [72] based on Euler–Bernoulli's beam theory. Subsequently, it was found that the vibration modes of DWCNTs are noncoaxial intertube vibrations and deflection of inner and outer nanotube can come about in same or opposite deflections.

More-over the vibration analysis of MWCNTs were implemented by Aydogdu [73] using generalized shear deformation beam theory (GSDBT). Parabolic shear deformation theory (PSDT) was used in the specific solutions and the results showed remarkable difference between PSDT and Euler beam theory and also the importance of vdW force presence for small inner radius. Lei et al. [74] have presented a theoretical vibration analysis of the radial breathing mode (RBM) of DWCNTs subjected to pressure based on elastic continuum model. It was shown that the frequency of RBM increases perspicuously as the pressure increases under different conditions.

The influences of shear deformation, boundary conditions and vdW coefficient on the transverse vibration of MWCNT were studied by Ambrosini and Borbón [75, 76]. The results reveal that the noncoaxial intertube frequencies are independent of the shear deformation and the boundary conditions and also are strongly influenced by the vdW coefficient used. The performed investigations on the simulation of vibrations properties classified by the method of modeling are presented in Tables 9.7–9.9. As it can be seen 5–7, 5–8, and 5–9, the majority of investigations used continuum modeling and simply replaced a CNT with hollow thin cylinder to study the vibrations of CNT. This modeling strategy cannot simulate the real behavior of CNT, since the lattice structure is neglected. In other word, these investigations simply studied the vibration behavior of a continuum cylinder with equivalent mechanical properties of CNT. Actually nano-scale continuum modeling is preferred for investigating vibrations of CNTs, since it was reported that natural frequencies of CNTs depend on both chirality and boundary conditions.

9.6 SUMMARY AND CONCLUSION ON CNT SIMULATION

The CNTs modeling techniques can be classified into three main categories of atomistic modeling, continuum modeling and nano-scale continuum modeling. The atomistic modeling consists of MD, MC and Ab initio methods. Both MD and MC methods are constructed on the basis of second Newton's law. While MD method deals with deterministic equations, MC is a stochastic approach. Although both MD and MC depend on potential, Ab initio is an accurate and potential free method relying on solving Schrödinger equation. Atomistic modeling techniques are suffering from some shortcomings which can be summarized as: (I) inapplicability of modeling large number of atoms (II) huge amount of computational tasks (III) complex formulations. Other atomistic methods such as tight bonding molecular dynamic, local density, density functional theory, Morse potential model and modified Morse potential model are also available which are in need of intensive calculations.

On the other hands, continuum modeling originated from continuum mechanics are also applied to study mechanical behavior of CNTs. Comprising of analytical and numerical approaches, the validity of continuum modeling has to be carefully observed wherein lattice structure of CNT is replaced with a continuum medium. Numerical continuum modeling is accomplished through finite element modeling using shell or curved plate elements. The degree to which this strategy, that is, neglecting lattice structure of CNT, will lead us to sufficiently accurate results is under

TABLE 9.7 MD Methods for Prediction Vibrational Properties of CNT

Researchers	Year	Method	Chairality	Aspect ratio (L/D)	Results
Gupta and Batra [77]	2008	MM3 potential	Nineteen armchair, zigzag, and chiral SWCNTs have been discussed	15	Axial, torsion and radial breathing mode (RBM) vibrations of free–free unstressed SWCNTs and identifying equivalent continuum structure
Gupta et al. [65]	2010	MM3 potential	Thirty-three armchair, zigzag band chiral SWCNTs have been discussed	3–15	Free vibrations of free end SWCNTs and identifying equivalent continuum structure of those SWCNTs
Ansari et al. [78]	2012	Adaptive Intermolecular Reactive Empirical Bond Order (AIREBO) potential	(8,8)	8.3–39.1	Vibration characteristics and comparison between different gradient theories as well as different beam assumptions in predicting the free vibrations of SWCNTs
Ansari et al. [79]	2012	Tersoff–Brenner and Lennard–Jones potential	(5,5)(10,10)(9,0)	3.61–14.46	Vibrations of single- and double-walled carbon nanotubes under various layer-wise boundary condition

TABLE 9.8 Continuum Methods for Prediction Vibration Properties of CNTs

Researchers	Year	Method	Chairality	Aspect ratio (L/D)	Results
Wang et al. [60]	2006	Timoshenko beam theory	0	10, 30, 50, 10	Free vibrations of MWCNTs using Timoshenko beam
Sun and Liu [80]	2007	Donnell's equilibrium equation	—	—	Vibration characteristics of MWCNTs with initial axial loading
Ke et al. [81]	2009	Eringen's nonlocal elasticity theory and von Karman geometric nonlinearity using nonlocal Timoshenko beam	—	10, 20, 30, 40	Nonlinear free vibration of embedded double-walled CNTs at different boundary condition
Ghavanloo and Fazelzadeh [82]	2012	Anisotropic elastic shell using Flugge shell theory	Twenty-four armchair, zigzag and chiral SWCNTs have been discussed	—	Investigating of free and forced vibration of SWCNTs including chirality effect
ydogdu [66]	2012	Nonlocal elasticity theory, an elastic rod model	—	—	Axial vibration of single walled carbon nanotube embedded in an elastic medium and investigating effect of various parameters like stiffness of elastic medium, boundary conditions and nonlocal parameters on the axial vibration
Khosrozadeh and Hajabasi [83]	2012	Nonlocal Euler–Bernoulli beam theory	—	—	Studying of nonlinear free vibration of DWCANTs considering nonlinear interlayer Van der Waals interactions, also discussing about nonlocal

TABLE 9.9 Nano-scale Continuum Modeling Simulation for Prediction Vibration Properties of CNTs

Researchers	Year	Method	Chairality	Aspect ratio (L/D)	Results
Georgantzinos et al. [84]	2008	Numerical analysis based on atomistic microstructure of nanotube by using linear spring element	Armchair and zigzag	Various aspect ratio	Investigating vibration analysis of single-walled carbon nanotubes, considering different support conditions and defects
Georgantzinos and Anifantis [85]	2009	FEM, based on linear nano-springs to simulate the interatomic behavior	Armchair and zigzag chiralities	3–20	Obtaining mode shapes and natural frequencies of MWCNTs as well as investigating the influence of van der Waals interactions on vibration characteristics at different BCs
Sakhaee Pour et al. [64]	2009	FEM using 3D beam element and point mass	Zigzag and armchair	Various aspect ratio	Computing natural frequencies of SWCNT using atomistic simulation approach with considering bridge and cantilever like boundary conditions

question. Moreover, it was extensively observed that almost all properties of CNTs (mechanical, buckling, vibrations and thermal properties) depend on the chirality of CNT; thus continuum modeling cannot address this important issue.

Recently, nano-scale continuum mechanics methods are developed as an efficient way of modeling CNT. These modeling techniques are not computationally intensive like atomistic modeling and thus they are able to be applied to more complex system with-out limitation of short time and/or length scales. Moreover, the discrete nature of the CNT lattice structure is kept in the modeling by replacing C–C bonds with a continuum element. Since the continuum modeling is employed at the scale of nano, therefore the modeling is called as nanoscale continuum modeling.

The performed investigations in literature addressing mechanical properties, buckling, vibrations and thermal behavior of CNT are reviewed and classified on the basis of three aforementioned modeling techniques. While atomistic modeling is a reasonable modeling technique for this purpose, its applicability is limited to the small systems. On the other hand, the continuum modeling neglects the discrete structure of CNT leading to inaccurate results.

Nano scale continuum modeling can be considered as an accept-able compromise in the modeling presenting results in a close agreement with than that of atomistic modeling. Employing FEM as a computationally powerful tool in nanoscale continuum modeling, the influence of CNT chirality, diameter, thickness and other involved parameters can be evaluated conveniently in comparison with other methods. Concerning CNTs buckling properties, many researchers have conducted the simulation using MD methods. The shell theories and molecular mechanic structure simulation are also applied to assess the CNTs buckling in order to avoid time consuming simulations. But, for the specific case of buckling behavior, it is not recommended to scarify the lattice structure of CNT for less time consuming computations. But instead, nanoscale continuum modeling is preferred.

Comparing the results with the obtained results of MD simulation, it can be inferred from literature that for large aspect ratios (i.e., length to diameter ratio $L/d > 10$) the simple Euler–Bernoulli beam is reliable to predict the buckling strains of CNTs while there fined Timoshenko's beam model or their nonlocal counterparts theory is needed for CNTs with intermediate aspect ratios (i.e., $8 < L/d < 10$). The Donnell thin shell theory is incapable to capture the length dependent critical strains for CNTs with small aspect ratios (i.e., L/d < 8). On the other hand, Sanders shell theory is accurate in predicting buckling strains and mode shapes of axially compressed CNTs with small aspect ratios. From the dynamic analysis point of view, the replacement of CNT with a hollow cylinder has to be extremely avoided, despite the widely employed method. In other word, replacing CNT with a hollow cylinder will not only lead us to inaccurate results, but also there will not be any difference between and nanostructure in the form of tube with continuum level of modeling. It is a great importance to keep the lattice structure of the CNT in the modeling, since the discrete structure of CNT play an important role in the dynamic analysis. From the vibration point of view, MD simulations are more reliable and NCM approaches are preferred to CM techniques using beam elements.

KEYWORDS

- **Carbon nanotubes (CNT)**
- **Computational chemistry**
- **Computational mechanics**

REFERENCES

1. Liew, K. M., He, X. Q. & Wong, C. W. (2004). On the Study of Elastic and Plastic Properties of Multi-Walled Carbon Nanotubes Under Axial Tension Using Molecular-dynamics Simulation. *Acta Mater*, *52,* 2521–2527.
2. Zhang, H. W., Wang, J. B. & Guo, X. (2005). Predicting the Elastic Properties of Single-Walled Carbon Nanotubes. *J Mech Phys Solids*, *53,* 1929–1950.
3. Agrawal Paras, M., Sudalayandi Bala, S., Raff Lionel, M. & Komanduri Ranga. (2006). Acomparison of Different Methods of Young's Modulus Determination For Single-Wall Carbon Nanotubes (SWCNT) Using Molecular Dynamics (MD)Simulations. *Comput. Mater. Sci., 38,* 271–281.
4. Cheng Hsien-Chie, Liu Yang-Lun, Hsu Yu-Chen, & Chen Wen-Hwa. (2009). Atomistic-Continuum Modeling for Mechanical Properties of Single-Walled Carbonnanotubes. *Int. J. Solids Struct., 46,* 1695–1704.
5. Cai, J., Wang, Y. D. & Wang, C. Y. (2009). Effect of Ending Surface on Energy and Young's Modulus of Single-Walled Carbon Nanotubes Studied Using Linear Scalingquantum Mechanical Method. *Physica B*, *404,* 3930–3934.

6. Ranjbartoreh Ali Reza, & Wang Guoxiu. (2010). Molecular Dynamic Investigation of Mechanical Properties of Armchair and Zigzag Double-Walled Carbonnanotubes Under Various Loading Conditions. *Phys. Lett. A, 374,* 969–974.
7. Sears, A. & Batra, R. C. (2004). Macroscopic Properties of Carbon Nanotubes from Molecular-Mechanics Simulations. *Phys. Rev. B, 69,* 235406.
8. Wang, Q. (2004). Effective In-Plane Stiffness and Bending Rigidity of Armchair and Zigzag Carbon Nanotubes. *Int. J. Solids Struct., 41,* 5451–5461.
9. Kalamkarov, A. L., Georgiades, A. V., Rokkam, S. K., Veedu, V. P. & Ghasemi-Nejhad, M. N. (2006). Analytical and Numerical Techniques to Predict Carbon Nanotubes Properties. *Int. J. Solids Struct., 43,* 6832–6854.
10. Gupta, S. S. & Batra, R. C. (2008). Continuum Structures Equivalent in Normal Mode-vibrations to Single-Walled Carbon Nanotubes. *Comput. Mater. Sci., 43,* 715–723.
11. Giannopoulos, G. I., Kakavas, P. A. & Anifantis, N. K. (2008). Evaluation of the Effective Mechanical Properties of Single-Walled Carbon Nanotubes Using a Spring Based Finite Element Approach. *Comput. Mater. Sci., 41(4),* 561–569.
12. Papanikos, P., Nikolopoulos, D. D. & Tserpes, K. I. (2008). Equivalent Beams for Carbonnanotubes. *Comput. Mater. Sci., 43,* 345–352.
13. Li Chunyu, & Chou Tsu-Wei. (2003). Elastic Moduli of Multi-Walled Carbon Nanotubes and the Effect of Van Der Waals Forces. *Compos. Sci. Technol., 63,* 1517–1524.
14. Natsuki, T., Tantrakarn, K. & Endo, M. (2004). Prediction of Elastic Properties for Single-Walled Carbon Nanotubes. *Carbon, 42,* 39–45.
15. Natsuki Toshiaki, & Endo Morinobu. (2004). Stress Simulation of Carbon Nanotubes in Tension and Compression. *Carbon, 42,* 2147–2151.
16. Xiao, J. R., Gama, B. A. & Gillespie, Jr J. W. (2005). An Analytical Molecular Structural Mechanics Model for the Mechanical Properties of Carbon Nanotubes. *Int. J. Solids Struct., 42,* 3075–3092.
17. Tserpes, K. I. & Papanikos, P. (2005). Finite Element Modeling of Single-Walled Carbon Nanotubes. *Composites: Part B, 36,* 468–477.
18. Jalalahmadi, B. & Naghdabadi, R. (2007). Finite Element Modeling of Single-Walled Carbon Nanotubes with Introducing a New Wall Thickness. *J. Phys.: Conf. Ser., 61,* 497–502.
19. Meo, M. & Rossi, M. (2006). Prediction of Young's Modulus of Single Wall Carbon Nanotubes by Molecularmechanics Based Finite Element Modeling. *Compos. Sci. Technol., 66,* 1597–1605.
20. PourAkbar Saffar Kaveh, JamilPour Nima, Najafi Ahmad Raeisi, RouhiGholamreza, Arshi Ahmad Reza, Fereidoon Abdolhossein, & et al. (2008). A Finite Element Model for Estimating Young's Modulus of Carbon Nanotube Reinforced Composites Incorporating Elastic Cross-links. *World Acad Sci.Eng Technol*, 47.
21. Cheng Hsien-Chie, Liu Yang-Lun, Hsu Yu-Chen, & Chen Wen-Hwa. (2009). Atomistic-Continuum Modeling for Mechanical Properties of Single-Walled Carbon Nanotubes. *Int. J. Solids Struct., 46,* 1695–1704.
22. Ávila Antonio Ferreira, & Lacerda Guilherme Silveira Rachid. (2008). Molecular Mechanics Applied to Single-Walled Carbon Nanotubes. *Mater Res, 11(3),* 325–333.
23. Wernik Jacob, M. & Meguid Shaker, A. (2010). Atomistic-Based Continuum Modeling of the Nonlinear Behavior of Carbon Nanotubes. *Acta Mech., 212,* 167–79

24. Shokrieh Mahmood, M. & Rafiee Roham. (2010). Prediction of Young's Modulus of Graphene Sheets and Carbon Nanotubes Using Nanoscale Continuum Mechanics Approach. *Mater Des*, *31*, 790–795.
25. Lu Xiaoxing, Hu Zhong. (2012). Mechanical Property Evaluation of Single-Walled Carbon Nanotubes by Finite Element Modeling. *Composites Part B*, *43*, 1902–1913.
26. Rafiee Roham, & Heidarhaei Meghdad. (2012). Investigation of Chirality and Diameter Effects on the Young's Modulus of Carbon Nanotubes Using Non-Linear Potentials. *Compos Struct.*, *94*, 2460–4.
27. Zhang, Y. Y., Tan, V. B. C. & Wang, C. M. (2006). Effect of Chirality on Buckling Behavior of Single-Walled carbon nanotubes. *J. Appl. Phys.*, *100*, 074304.
28. Chang, T. (2007). Torsional Behavior of Chiral Single-Walled Carbon Nanotubes is Loading Direction Dependent. *Appl. Phys. Lett.*, *90*, 201910.
29. Chang Tienchong, Li Guoqiang, & Guo Xingming. (2005). Elastic Axial Buckling of Carbonnanotubes Via a Molecular Mechanics Model. *Carbon*, *43*, 287–94.
30. Wang, C. M., Zhang, Y. Y., Xiang, Y. & Reddy, J. N. (2010). Recent Studies on Buckling of carbonnanotubes. *Appl Mech Rev*, *63*, 030804.
31. Srivastava, D., Menon, M. & Cho, K. J. (1999). Nanoplasticity of Single-Wall Carbon Nanotubes Under Uniaxial Compression. *Phys. Rev. Lett*, *83(15)*, 2973–2976.
32. Ni, B., Sinnott, S. B., Mikulski, P. T. & et al. (2002). Compression of Carbon Nanotubes Filled with C-60, CH4, or Ne: Predictions from Molecular Dynamics Simulations. *Phys. Rev. Lett*, *88*, 205505.
33. Wang, Yu., Wang Xiu-xi, Ni Xiang-gui, & Wu Heng-an. (2005). Simulation of the Elastic Response and the Buckling Modes of Single-Walled Carbon Nanotubes. *Comput. Mater. Sci.*, *32*, 141–146.
34. Hao Xin, Qiang Han, & Xiaohu Yao. (2008). Buckling of Defective Single-Walled and Double-Walled Carbon Nanotubes Under Axial Compression by Moleculardynamics Simulation. *Compos. Sci. Technol.*, *68*, 1809–1814.
35. Ansari, R., Sahmani, S. & Rouhi, H. (2011). Rayleigh–Ritz Axial Buckling Analysis of Single-Walled Carbon Nanotubes with Different Boundary Conditions. *Phys. Lett. A*, *375*, 1255–1263.
36. Yakobson, B. I., Brabec, C. J. & Bernholc, J. (1996). Nanomechanics of Carbon Tubes: Instabilities Beyond Linear Response. *Phys. Rev. Lett*, *76*, 2511–2514.
37. Zhang Chen-Li, & Shen Hui-Shen. (2006). Buckling and Postbuckling Analysis of Single-Walled Carbon Nanotubes in Thermal Environments Via Molecular Dynamics Simulation. *Carbon*, *44*, 2608–2616.
38. Jeong Byeong-Woo, & Sinnott Susan, B. (2010). Unique Buckling Responses of Multi-Walled Carbon Nanotubes Incorporated as Torsion Springs. *Carbon*, *48*, 1697–1701.
39. Silvestre, N., Wang, C. M., Zhang, Y. Y. & Xiang, Y. (2011). Sanders Shell Model for Buckling of Single-Walled Carbon Nanotubes with Small Aspect Ratio. *Compos Struct.*, *93*, 1683–1691.
40. Donnell, L. H. (1933). Stability of Thin-Walled Tubes Under Torsion, NACA Report No. 479.
41. Sanders, J. L. (1963). Non-Linear Theories for Thin Shells. *Quart J Appl Math*, *21*, 21–36.

42. Silvestre Nuno, Faria Bruno, & Canongia Lopes José N. (2012). A Molecular Dynamics Study on the Thickness and Post-Critical Strength of Carbon Nanotubes. *Compos Struct., 94,* 1352–1358.
43. Ghavamian Ali, & Öchsner Andreas. (2012). Numerical Investigation on the Influence of Defects on the Buckling Behavior of Single-and Multi-Walled Carbon *Nanotubes. Physica E, 46,* 241–249.
44. Zhang, Y. Y., Wang, C. M., Duan, W. H., Xiang, Y. & Zong, Z. (2009). Assessment of Continuummechanics Models in Predicting Buckling Strains of Single-Walled Carbonnanotubes. *Nanotechnology, 20,* 395707.
45. Sears, A. & Batra, R. C. (2006). Buckling of Multiwalled Carbon Nanotubes Under Axial Compression. *Phys. Rev. B, 73,* 085410.
46. Xin Hao, Han Qiang, & Yao Xiao-Hu. (2007). Buckling and Axially Compressive Properties of Perfect and Defective Single-Walled Carbon Nanotubes. *Carbon, 45,* 2486–2495.
47. Silvestre Nuno, Faria Bruno, & Canongia Lopes José N. (2012). A Molecular Dynamics Study on the Thickness and Post-Critical Strength of Carbon Nanotubes. *Compos Struct., 94,* 1352–1358.
48. He, X. Q., Kitipornchai, S. & Liew, K. M. (2005). Buckling analysis of Multi-Walled Carbon Nanotubes: a Continuum Model Accounting for Van Der Waals Interaction. *J. Mech Phys Solid, 53,* 303–326.
49. Ghorbanpour Arani, A., Rahmani, R. & Arefmanesh, A. (2008). Elastic Buckling Analysis of Single-Walled Carbon Nanotube Under Combined Loading by Using the ANSYS Software. *Physica E, 40,* 2390–2395.
50. Yao Xiaohu, Han Qiang, & Xin Hao. (2008). Bending Buckling Behaviors of Single- and Multi-Walled Carbon Nanotubes. *Comput. Mater. Sci., 43,* 579–590.
51. Guo, X., Leung, A. Y. T., He, X. Q., Jiang, H. & Huang, Y. (2008). Bending Buckling of Single-Walled Carbon Nanotubes by Atomic-Scale Finite Element. *Composites: Part B, 39,* 202–208.
52. Chan Yue, Thamwattana Ngamta, & Hill James, M. (2011). Axial Buckling of Multi-Walled Carbon Nanotubes and Nanopeapods. *Eur. J. Mech. A/Solid, 30,* 794–806.
53. Silvestre Nuno. (2012). On the Accuracy of Shell Models for Torsional Buckling of Carbon Nanotubes. *Eur. J. Mech. A/Solid, 32,* 103–108.
54. Li Chunyu, & Chou Tsu-Wei. (2004). Modeling of Elastic Buckling of Carbon Nanotubes by Molecular Structural Mechanics Approach. *Mech Mater, 36,* 1047–1055.
55. Hu, N., Nunoya, K., Pan, D., Okabe, T. & Fukunaga, H. (2007). Prediction of Buckling Characteristics of Carbon Nanotubes. *Int. J. Solids Struct., 44,* 6535–6550.
56. Sawano, S., Arie, T. & Akita, S. (2010). Carbon Nanotube Resonator in Liquid. *Nano Lett, 10,* 3395–3398.
57. Formica Giovanni, Lacarbonara Walter, & Alessi Roberto. (2010). Vibrations of Carbonnanotube-Reinforced Composites. *J. Sound Vib., 329,* 1875–1889.
58. Khan Shafi Ullah, Li Chi Yin, Siddiqui Naveed A. & Kim Jang-Kyo. (2011). Vibration Damping Characteristics of Carbon Fiber-Reinforced Composites Containing Multi-Walled Carbon Nanotubes. *Compos. Sci. Technol., 71,* 1486–1494.
59. Gibson, R. F., Ayorinde, E. O. & Wen, Y. F. (2007). Vibrations of Carbon Nanotubes and their Composites: a Review. *Compos. Sci. Technol., 67,* 1–28.

60. Wang, C. M., Tan, V. B. C. & Zhang, Y. Y. (2006). Timoshenko Beam Model for Vibration Analysis of Multi-Walled Carbon Nanotubes. *J. Sound Vib., 294,* 1060–1072.
61. Hu Yan-Gao, Liew, K. M. & Wang, Q. (2012). Modeling of Vibrations of Carbon Nanotubes. *Proc. Eng., 31,* 343–347.
62. Li, C. & Chou, T. W. (2003). Single-Walled Carbon Nanotubes as Ultrahigh Frequency Nanomechanical Resonators. *Phys. Rev. B, 68,* 073405.
63. Arghavan, S. & Singh, A. V. (2011). On the Vibrations of Single-Walled Carbon Nanotubes. *J. Sound Vib., 330,* 3102–3122.
64. Sakhaee-Pour, A., Ahmadian, M. T. & Vafai, A. (2009). Vibrational Analysis of Single-Walled Carbon Nanotubes Using Beam Element. *Thin Wall Struct., 47,* 646–652.
65. Gupta, S. S., Bosco, F. G. & Batra, R. C. (2010). Wall Thickness and Elastic Moduli of Single-Walled Carbon Nanotubes from Frequencies of Axial, Torsional and in Extensional Modes of Vibration. *Comput. Mater. Sci., 47,* 1049–1059.
66. Aydogdu Metin. (2012). Axial Vibration Analysis of Nanorods (carbon nanotubes) Embedded in an Elastic Medium Using Nonlocal Elasticity. *Mech Res Commun., 43,* 34–40.
67. Fakhrabadi Mir Masoud Seyyed, Amini Ali, & Rastgoo Abbas. (2012). Vibrational Properties of Two and Three Junctioned Carbon Nanotubes. *Comput. Mater. Sci., 65,* 411–425.
68. Lee, J. H. & Lee, B. S. (2012). Modal Analysis of Carbon Nanotubes and Nanocones Using FEM. *Comput. Mater. Sci., 51,* 30–42.
69. Fu, Y. M., Hong, J. W. & Wang, X. Q. (2006). Analysis of Nonlinear Vibration for Embedded Carbon Nanotubes. *J. Sound Vib., 296,* 746–756.
70. Ansari, R. & Hemmatnezhad, M. (2011). Nonlinear Vibrations of Embedded Multi-Walled Carbon Nanotubes Using a Variational Approach. *Math Comput Modell, 53,* 927–938.
71. Wang, C. Y., Li, C. F. & Adhikari, S. (2010). Axisymmetric Vibration of Single-Walled Carbonnanotubes in Water. *Phys. Lett. A, 374,* 2467–2474.
72. Natsuki Toshiaki, Ni Qing-Qing, & Endo Morinobu. (2008). Analysis of the Vibration Characteristics of Double-Walled Carbon Nanotubes. *Carbon, 46,* 1570–1573.
73. Aydogdu Metin. (2008). Vibration of Multi-Walled Carbon Nanotubes by Generalized Shear Deformation Theory. *Int. J. Mech. Sci., 50,* 837–844.
74. Lei Xiao-Wen, Natsuki Toshiaki, Shi Jin-Xing, & Ni Qing-Qing. (2011). Radial Breathing Vibration of Double-Walled Carbon Nanotubes Subjected to Pressure. *Phys. Lett. A, 375,* 2416–2421.
75. Ambrosini Daniel, & de Borbón Fernanda. (2012). On the Influence of the Shear Deformation and Boundary Conditions on the Transverse Vibration of Multi-Walled Carbon Nanotubes. *Comput. Mater. Sci., 53,* 214–219.
76. de Borbón Fernanda, & Ambrosini Daniel. (2012). On the Influence of Van Der Waals Coefficient on the Transverse Vibration of Double Walled Carbon Nanotubes. *Comput. Mater. Sci., 65,* 504–508.
77. Gupta, S. S. & Batra, R. C. (2008). Continuum Structures Equivalent in Normal Mode-vibrations to Single-Walled Carbon Nanotubes. *Comput. Mater. Sci., 43,* 715–723.
78. Ansari, R., Gholami, R. & Rouhi, H. (2012). Vibration Analysis of Single-Walled Carbon Nanotubes Using Different Gradient Elasticity Theories. *Composites: Part B, 43(8),* 2985–2989.

79. Ansari, R., Ajori, S. & Arash, B. (2012). Vibrations of Single- and Double-Walled Carbonnanotubes with Layer-Wise Boundary Conditions: a Molecular Dynamics Study. *Curr. Appl. Phys., 12,* 707–711.
80. Sun, C. & Liu, K. (2007). Vibration of Multi-Walled Carbon Nanotubes with Initial Axialloading. *Solid State Commun., 143,* 202–207.
81. Ke, L. L., Xiang, Y., Yang, J. & Kitipornchai, S. (2009). Nonlinear Free Vibration of Embedded Double-Walled Carbon Nanotubes Based on Nonlocal Timoshenko Beam Theory. *Comput. Mater. Sci., 47,* 409–417.
82. Ghavanloo, E. & Fazelzadeh, S. A. (2012). Vibration Characteristics of Single-Walled Carbonnanotubes Based on an Anisotropic Elastic Shell Model Including Chirality Effect. *Appl. Math. Model, 36,* 4988–5000.
83. Khosrozadeh, A. & Hajabasi, M. A. (2012). Free Vibration of Embedded Double-Walled Carbon Nanotubes Considering Nonlinear Interlayer Van Der Waals Forces. *Appl. Math. Model, 36,* 997–1007.
84. Georgantzinos, S. K., Giannopoulos, G. I. & Anifantis, N. K. (2009). An Efficient Numerical Model for Vibration Analysis of Single-Walled Carbon Nanotubes. *Comput. Mech., 43,* 731–741.
85. Georgantzinos, S. K. & Anifantis, N. K. (2009). Vibration Analysis of Multi-Walled Carbonnanotubes Using a Spring–Mass Based Finite Element model. *Comput. Mater. Sci., 47,* 168–177.

INDEX

B

C

G

H

I

N

O

P

V

W

X

Y

Z